CHEMICAL FORMULATION
An Overview of Surfactant-based Preparations Used in
Everyday Life

RSC Paperbacks

RSC Paperbacks are a series of inexpensive texts suitable for teachers and students and give a clear, readable introduction to selected topics in chemistry. They should also appeal to the general chemist. For further information on all available titles contact:

Sales and Customer Care Department, Royal Society of Chemistry,
Thomas Graham House, Science Park, Milton Road, Cambridge CB4 0WF, UK
Telephone: +44 (0)1223 432360; Fax: +44 (0)1223 423429; E-mail: sales@rsc.org

Recent Titles Available

The Chemistry of Fireworks
By Michael S. Russell
Water (Second Edition): A Matrix of Life
By Felix Franks
The Science of Chocolate
By Stephen T. Beckett
The Science of Sugar Confectionery
By W.P. Edwards
Colour Chemistry
By R.M. Christie
Beer: Quality, Safety and Nutritional Aspects
By P.S. Hughes and E.D. Baxter
Understanding Batteries
By Ronald M. Dell and David A.J. Rand
Principles of Thermal Analysis and Calorimetry
Edited by P.J. Haines
Food: The Chemistry of Its Components (Fourth Edition)
By Tom P. Coultate
The Misuse of Drugs Act: A Guide for Forensic Scientists
L.A. King

Future titles may be obtained immediately on publication by placing a standing order for RSC Paperbacks. Information on this is available from the address above.

RSC Paperbacks

CHEMICAL FORMULATION
An Overview of Surfactant-based Preparations Used in Everyday Life

TONY HARGREAVES

advancing the chemical sciences

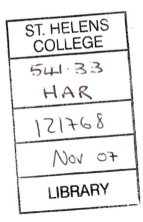
ISBN 0–85404–635–6

A catalogue record for this book is available from the British Library

Published by The Royal Society of Chemistry,
Thomas Graham House, Science Park, Milton Road,
Cambridge CB4 0WF, UK

For further information see our web site at www.rsc.org

Typeset in Great Britain by RefineCatch Ltd, Bungay, Suffolk, UK
Printed by Cambridge University Press, Cambridge, UK

Preface

My first experience of the chemistry of surfactant-based formulations was when I was working on analysis of solvent blends by Gas Chromatography (GC) back in the early 1980s. In addition to GC work our laboratory would also carry out comparative studies of surfactant formulations like shampoos and related products.

Instrumental methods such as GC or Infra-red (IR) were out of the question because of the disproportionate amount of preparation required to make the sample, or a part of it, 'instrument ready'. As such many of these comparative studies were confined to quick and simple tests: pH, conductivity, solids, ash and indicator tests for surfactant type.

From time to time we would be asked: Can you test this and give an idea of how to make an equivalent? Although cautious of trespassing on someone else's intellectual property we would often take up the challenge of making and matching formulations based on surfactants. It was an interesting diversion: instead of the analytical approach of breaking down formulations we could go in the opposite direction and build them up.

It was then that I began to appreciate the difficulty of getting even a basic understanding of what goes into formulations, the amounts used and how each component functions. For a start the raw materials were commercial products with commercial names and often it was a problem knowing what the chemistry was. Although there were many formulations provided by chemical suppliers for us to study there seemed to be few, if any, guidance rules. It was only after a lengthy exercise of studying formulations, basic surfactant chemistry and discussions with surfactant manufacturers that I was able to form a generalised picture of what was going on.

This book is a summary of the chemistry and technology that was pulled together in the attempts to see more clearly what chemicals go into everyday preparations, how those chemicals function and some related aspects such as environmental effects. Hopefully the book will be of value for those with an interest in household chemicals, others who may wish to consider making such products as a commercial venture and students with an interest in applied chemistry.

In preparing this book I have used many formulations provided by surfactant manufacturers and to them I wish to express my gratitude. In particular I wish to thank Nick Challoner and Rebecca Webster at Croda Oleochemicals for permission to use their formulations and for the many helpful suggestions to ensure that what is reported is as up-to-date and accurate as possible. I would wish to extend my thanks to Chris Howard at Stepan UK and to Simon Nicholson at Huntsman Surface Sciences for their kind assistance.

Contents

Contents

Abbreviations Used

AG   Alkylpolyglucoside
ASTM   American Society for Testing Materials
BHA   Butylated hydroxyanisole
BHT   Butylated hydroxytoluene
BOD   Biochemical oxygen demand
BS   British Standard
CAS   Chemical Abstracts Service
CDE   Coconut diethanolamide
CFC   Chlorofluorocarbon
CI   Colour index
CMC   Critical micelle concentration
COD   Chemical oxygen demand
CoSHH   Control of Substances Hazardous to Health
CTFA   Cosmetics Toiletries and Fragrance Association
DEA   Diethanolamine
DSP   Disodium phosphate
DTPA   Diethylenetriaminepentaacetate
EDTA   Ethylenediaminetetraacetate
EO   Ethylene oxide
GC   Gas chromatography
GWP   Global warming potential
HCFC   Hydrochlorofluorocarbon
HEDTA   Hydroxyethylethylenediaminetriacetate
HFC   Hydrofluorocarbon
HFE   Hydrofluoroether
HLB   Hydrophile lipophile balance
HPLC   High performance liquid chromatography
INCI   International Nomenclature of Cosmetic Ingredients
IPA   Isopropyl alcohol
IR   Infra-red
ISO   International Organization for Standardization
IUPAC   International Union of Pure and Applied Chemists

LABS	Linear alkylbenzene sulfonate
LCA	Life cycle assessment
LCI	Life cycle inventory
MALDI	Matrix Assisted Laser Desorption Ionization
MEA	Monoethanolamine
MS	Mass spectrometry
MSDS	Material Safety Data Sheet
MSP	Monosodium phosphate
NMR	Nuclear magnetic resonance
NTA	Nitrilotriacetate
o/w	Oil-in-water (emulsion)
ODP	Ozone depletion potential
PEG	Polyethylene glycol
PO	Propylene oxide
PVC	Polyvinyl chloride
SCMC	Sodium carboxymethylcellulose
STPP	Sodium tripolyphosphate
TEA	Triethanolamine
TKPP	Tetrapotassium pyrophosphate
TOFA	Tall oil fatty acids
uPVC	Unplasticized polyvinyl chloride
UV	Ultra-violet
VOC	Volatile organic compound
w/o	Water-in-oil (emulsion)

Introduction

Chemical formulations go so far back in time that we must turn to the archaeologist for examples. As long ago as Mesolithic times when people were hunter-gatherers they attached worked flint arrow heads to shafts by means of a blend of resin extracted from birch bark and made plastic with beeswax to produce a substance that had the best properties for the job.

Resin alone, once set, produces a brittle mass but when blended with just the right amount of beeswax it provides not only a strong adhesive but also has sufficient flexibility to withstand the mechanical forces imposed on a weapon such as an arrow.

Found only in certain limestone deposits flint was scarce and often was brought from far and wide. And so it seems reasonable to assume that a great deal of care went into any experiments of which it formed a part. Quite simply, a lost arrow point demanded a replacement that took a skilled person some time to make and required a fairly large chunk of that valuable material.

One can imagine that the process would follow the lines: theory, experiment, result, interpretation, new theory, new experiment, new result and so on until performance was optimized. Those stone-age experiments were in fact an early application of plasticizing a polymer – the beeswax was the plasticizer. The product must have been used as a hot melt adhesive which, on cooling, left the flint arrowhead firmly, but flexibly, bonded to its pine shaft.

The very same principle of plasticizing a polymer is used today to turn rigid uPVC (unplasticized PVC) into pliable PVC. Yes, we use a ten thousand year old discovery to take the 'u' out of uPVC. There are other examples from antiquity but the beeswax/resin example is of particular significance in a scientific context as it would have called for a good deal of quantitative experimental work to optimize the properties of the blend.

1

Eventually the hunter-gatherer lifestyle gave way to living in settled communities as a result of early agriculture. Now, instead of travelling far and wide to catch animals the task was simplified with much time and effort saved. With the animals in their pens and people living in settled communities two very favourable conditions, the availability of spare time and permanent work space, provided for a basic knowledge of materials to be developed.

The interest in materials led to the discovery of many chemical reactions that were to mark the onset of materials technology. Thus the foundation was laid for the high temperature chemistry that turns sand and clay into pot. The importance of recognizing essential properties was being learnt: quality of raw materials, proportions to blend together to achieve certain effects, estimation of kiln temperature, rate and duration of firing.

Later on, other high temperature reaction based processes were discovered, studied and refined to give rise to a new technology based on chemical reduction of a greenish rock to produce bronze. Still later, the iron-age, based on turning brown iron ore into iron developed. Thus, chemical technology was well established in times of antiquity. Much of the chemistry involved was brute force chemical reaction, high temperature stuff.

In general, chemistry is recognized in its ability to bring about reactions to effect a permanent change from one substance to another; the latter bearing no resemblance to that from which it was made – just like the green rock to bronze referred to above. However, the chemistry involved in making the plasticized resin did not involve chemical reactions to create new substances but was the practice of making chemical mixtures. It is the preparation of mixtures that is the essence of formulation work.

The properties of mixtures reflect the properties of the substances from which they are made and the mixtures can, by means of simple techniques, be split into their component substances. Often, to the frustration of the chemical formulator, a prepared mixture separates out on its own.

So, now to consider some simple two-part mixtures; some of these are natural and others are man-made ones:

Components	Type of mixture	Example
Solid in solid	Powder blend	Facial make-up powder
	Solid solution	Bronze
Liquid in liquid	Solution	Alcohol in water
	Emulsion	Mayonnaise

(Continued)

Components	Type of mixture	Example
Gas in gas	Perfect mixture	Air
Solid in gas	Smoke	Cigarette smoke
Gas in solid	Solid foam	Pumice stone
Liquid in gas	Aerosol	Fog
Gas in liquid	Foam	Shaving foam
	Solution	Sparkling water
Solid in liquid	Solution	Salt water
	Suspension	Milk of magnesia
Liquid in solid	Solid solution	Mercury gold amalgam
	Paste	Clay

Some everyday materials, particularly those that form parts of living systems, go even further and involve combinations of the above types in some incredibly complex mixtures. In the formulations discussed in the following pages the mixtures are relatively simple: solutions, emulsions, suspensions and foams. But just because a mixture is simple in terms of the number of components does not mean it is simple at the molecular level.

The aim in making formulated preparations is to take readily available substances (man-made or natural) to arrive at a preparation that is tailor-made for a specific task, a task that cannot be effectively carried out by a single substance working on its own. This requires a good deal of applied science and much of that is chemistry. In addition to the technical aspects of putting a preparation together there will also have been a study of many of the other aspects that are essential if a preparation is to be a commercial product.

Each of the following all have a role to play: formulation recipe, chemicals information, reactivity considerations, blending method, testing and analysis, stability trials (physical/chemical/microbiological) and shelf-life, health and safety, environment resources and waste, quality control, packaging and container interactions, labelling and user instructions and warnings as to adverse effects.

In all of this there is a big responsibility to ensure the product is fit for its intended purpose. This extends to an awareness of how the chemical blend may interact, physically or chemically, with the materials with which it comes into contact during use.

For example, a cleaning preparation based on ammonia may well have excellent performance on most surfaces but failure to point out that it should not be used on marble surfaces could turn out to be embarrassing, as could the fact that an acid-based metal brightener when in contact with stainless steel turns the steel black. Of course not all situations can be predicted but what happens to a preparation once it is in the user's hands is something for consideration.

There is a tendency to think that once a formulation has been put together and shown to be stable, over a period of days, that it will remain as such *ad infinitum*. This is almost never the case. It is not unknown for containers with apparently stable, safe and unreactive mixtures in them, even after sitting in a cold storeroom for months or years, to explode due to gassing of the contents or for a natural solvent blend, after being put into a polythene bottle, to ooze out through the polythene and leave a crumpled and collapsed bottle sitting in a sticky pond.

Chemical preparations are the basis of a whole range of modern materials in the home and at work. By far the largest proportion of these – largest in terms of both tonnage and variety – is formulations that involve surfactants. In fact it was developments in surfactant chemistry that laid the foundation for many of the chemical formulations we see today.

Surfactants are to be found as important ingredients in shampoos, shower gel, skin creams and ointments, bactericides, surface cleaners, polishes and abrasives, washing powders and liquids, fabric conditioners, engine oil, food and drink, pharmaceuticals, agricultural pesticides, emulsion paints, adhesives, bleach, solvents, toothpastes and mouthwash. The list is endless because it continues to grow.

Today much of the momentum of that growth comes from our desire to move away from solvent preparations, such as the volatile organic compounds (VOCs) that pollute the atmosphere, towards water-based systems, and that means surfactant-based ones.

Any chemical preparation must remain as a coherent mixture for a reasonable length of time; it must be stable. A product with poor stability in which the components separate out and which requires an expiry date only two weeks from date of manufacture is not going to go far in terms of customer satisfaction.

Achieving what at first sight seem to be simple criteria is hardly ever straightforward. For example, a solution of gum acacia is easy to prepare and, to begin with, gives all the impression of being stable. However, a few days later the situation will be very different as it has turned cloudy and clumps of fungal growth adorn its surface.

What seemed to be a simple task of weighing out, dissolving, straining and bottling is now a bit more involved. Back to the drawing board and time to consider what preservatives are available, how compatible they are and what concentrations they are to be used at.

Planning a chemical formulation is not an exact science. To set about applying theoretical principles and chemical data, doing calculations or computer modelling to examine equilibria for, say, five chemicals mixed together and distributed between two phases is, if not downright impossible, certainly unrealistic. Much of the data required for this type of exercise is just not available – there are parameters that no one has measured.

Thus, many chemical preparations are put together on the basis of some general chemical theory, an element of creativity and a lot of trial and failure experimentation. Recently, software packages have become available which can be used to design an experimental formulation and these can streamline the whole process by reducing the number of experiments. They also allow for 'what if' exercises to be carried out. Despite the introduction of such tools many formulators still view their task as being as much of an art as a science.

In practice we find that the bulk of formulation work is the business of a few large, well-established, companies. But as with all business there are a greater number of smaller ones and an even greater number of embryonic ones. Many of these embryos are at risk of spontaneous abortion once their entrepreneurs – usually non-chemists – realize that to make a liquid for 30p per litre and sell it for £3 per litre requires a little more in the way of know how and effort than simply throwing a load of chemicals into a plastic drum, topping up with water, stirring for a few minutes with the broom handle and bottling it.

When such an approach is used in the preparation of surfactant blends it is not uncommon for the mixture to form slimy lumps that positively refuse to dissolve and bob about on the surface. A mix such as this is destined to be sealed into a drum, labelled 'quarantine' and kept in the back yard because its creator is clueless as to how to get rid of it and his colleague thinks it may come in useful one day – perhaps lumps of slime will be all the rage next year.

People with no knowledge of formulation work getting in on the business of mixing chemicals can certainly be a problem. Even those with considerable experience can get themselves into a mess so what chance does the novice stand?

The fact that it seems a simple task to mix a few chemicals together is probably one of the attractions of the formulation business. Consequently, formulation work represents the do-it-yourself end of the chemical industry and is often regarded as the poor relative of a very respectable family. Sadly, it has attracted a few people who, with no technical knowledge at all – or even any regard for it – go out and buy a few chemicals, mix them together on a wing and a prayer and sell the resulting concoction to the public.

For those with a basic knowledge of applied chemistry the job of preparing good, stable, safe and effective formulations should not be too daunting once some groundwork has been completed. The problem is: where do potential formulators get the basic information from? Certainly the chemical suppliers will be only too pleased to provide recipes for preparations based on their own chemicals.

These are excellent sources of information but the very fact that it comes from a particular supplier wanting to promote – quite naturally – his/her own chemicals as raw materials, imposes restrictions. A problem here is that these chemicals will be sold under a trade name and so its chemical composition may not be entirely clear. Thus, comparison with chemicals from other suppliers is made that much more difficult.

So, where else to look for information? The potential formulator may be tempted to do a literature or web search to see if anyone has published the results of their research on how to make the world's best selling shampoo. Although in many areas of chemistry the results of research and development are published, much of the knowledge about commercial formulations is kept in private collections and is certainly not offered for all and sundry to learn about. Clearly, a company that has invested much time and money having its formulation team develop a recipe for a car wash-and-wax preparation that is going to knock the competition for a six is hardly likely to shout its recipe from the roof tops.

The aim of this book is to go some way to filling the information gap between textbook chemistry and the applied chemistry of everyday chemical formulations. Some pre-knowledge of chemistry basics is required to understand the chemical technology that is the foundation for those everyday products.

Chapter 1

Formulation Chemicals

Over the past few years the word chemical has taken on some very negative connotations. Apart from being difficult to define, the word all too often generates the wrong images in people's minds and is the source of the current chemophobia.

There are some among the anti-chemicals lobby who create confusion by promoting the idea that chemicals are nasty man-made substances, damaging to humans, animals and the environment and that we should use natural materials that are both healthy and environmentally friendly.

Such messages are unhelpful and foster in the minds of the non-scientists notions that there are natural substances and there are chemicals and the two are totally different things. This is unfortunate but we might expect those with only a basic knowledge of chemistry to see the flaw in it. Sadly this is not the case for even those with a scientific background can be confused.

I asked some A Level chemistry students: What is a chemical? There was an instant response that went something like this: 'Dangerous substances like acids, poisonous gases, pesticides and other stuff that damages the environment'. So, where do we go from there?

It is my view that any text on chemicals or applied chemistry has a duty to acknowledge those fears, put things in a true perspective and hopefully bring about some reassurance. And to do this it is essential to be absolutely clear as to what a chemical is.

Even if this goes only as far as explaining that the whole of the material world is atoms, produced by nature billions of years ago, chemically bonded together to produce the millions of molecules that are us and our world. It does not matter whether the bonding together of those atoms was done by nature or by man.

Good old alcohol is a prime example. The ethanol molecules produced from sugars by the catalytic effect of enzymes in yeast cells, a natural process, are indistinguishable from ethanol molecules made by man in the catalytic reaction between steam and ethene (ethylene). The water molecules exhaled by us as one of the waste products of our metabolizing carbohydrates, fats and proteins are no different from those exhaled by our beloved cars as they gobble up hydrocarbons in their infernal combustion engines.

It has to be accepted that not all man-made chemicals are a good thing just as not all nature-made chemicals are a good thing. Nature's chemical factory can come up with some treacherous substances capable of horrendous damage to the environment and living systems.

Some man-made chemicals have no analogues in nature and these are the ones that need to be watched carefully as, once they are done with and despatched to the environment, they do not always fit happily into nature's recycling mechanisms. The chlorofluorocarbons (CFCs) that damage the ozone layer are a recent example of this. Many of those molecules that are man-made, but are not straight copies of nature-made ones, we would not like to be without. Take for example the modern anaesthetics. What sort of surgery could be carried out without them?

We have gone a long way down the road to synthetic living. In fact our very bodies are now partly man-made. Those protein molecules that form a major part of us have nitrogen and hydrogen with chemical bonds that were made in chemical works in the Haber synthesis of ammonia. The ammonia so made is then used to make fertilizers which then make plant protein which we eat – directly or *via* animals – and incorporate into our tissue. Yes, we are partly synthetic.

For the purpose of this text I shall regard a chemical simply as a single substance (element or compound) with a definite and fixed composition that can be expressed precisely by means of a formula whether or not the substance is natural or man-made.

In classifying a chemical as natural, man-made or somewhere in between there is a need to examine its origin and any changes it has been subjected to before it is put to use. Take for example gum arabic, a traditional thickener in chemical formulations, which is entirely natural. The lumps of resinous material are a carbohydrate gum that is exactly as produced by the acacia tree and with no chemical modification. Man has simply collected it, picked out some contaminating bits of bark and despatched it to the user. The product is 100% natural.

At the other extreme there is tetrachloromethane, CCl_4. This is as man-made as you can get, not one chemical bond from a natural substance exists. Most chemicals fall somewhere in between these extremes in that they retain at least some of their nature-made structure. An example of one of these 'in between' compounds is sodium oleate,

$CH_3(CH_2)_7CH:CH(CH_2)_7COONa$, a type of soap made from vegetable oil and sodium hydroxide, NaOH. In the molecule the carbon chain, $CH_3(CH_2)_7CH:CH(CH_2)_7CO$, is unaltered; it is the ONa that is due to our chemical technology.

The amount of natural content can be quantified by means of simple calculations based on molecular mass. For sodium oleate the natural component amounts to 87%. This is a highly simplified way of assessing substances and no regard is made of the other factors, like the production of the sodium hydroxide, that are essential to the manufacturing process.

It may be that a molecule comes out at over 80% nature-made content but its 20% man-made part involved so many other chemicals that it fares less well than a molecule of a mere 10% nature-made content. This is to say nothing of the energy aspects of the environment/resource equation. The foregoing may be a rather crude instrument but it is a useful exercise in our efforts to focus upon more natural chemicals; it will also be helpful when we consider 'cradle-to-grave' analysis in Chapter 5.

Some chemicals used in formulations are petroleum based and are obtained simply by separation processes such as distillation. As such they retain their natural structure and in this respect they are natural substances but, in contrast to the oleate (above) which is made from vegetable oil, they are not from renewable resources.

In considering the worthiness of a chemical as an ingredient in a preparation the chemist of today must at least pay some attention as to whether it is from renewable or non-renewable resources. Unfortunately, however hard we work to avoid the use of petroleum it will still play a part somewhere in the process – at least for the foreseeable future.

In the discussions on chemicals that follow there will be frequent reference to some relevant chemistry theory. This has been kept to the minimum required for an understanding of the types of formulations discussed in this book.

CHEMICAL NAMES

There are over one hundred thousand chemicals currently in use but this represents only a small fraction of those that are known. In fact, the total number of chemical substances registered with the Chemicals Abstract Service is over 18 million and new ones arrive at a rate of over 300 per hour. With this number of different chemicals, and the need for each one to be completely distinguishable from the rest of the crowd, comes a problem of naming.

Systematic approaches for naming chemical compounds have to be

used but, all too frequently, this presents us with names that are virtually
unusable in written and spoken communications. Try this one in discussion – and it's by no means one of the complicated ones – *N*,*N*′-1,2-
ethanediylbis[*N*-(carboxymethyl)glycine]tetrasodium; it is usually called
EDTA tetrasodium. So where does this abbreviation EDTA come from?
That is from an earlier naming system: ethylenediaminetetraacetic acid.
Unfortunately for the chemist of today there are frequently many names
for the same substance.

A look in the *Merck Index* for our EDTA, which puts it under the
heading Edetate Sodium, gives many other names. In addition to systematic names there are trivial names, common names and, if that were
not complex enough, commerce comes in with a whole string of trade
names. Thus, for the EDTA we have, from the *Merck Index*: ethylene
bis(iminodiacetic acid) disodium salt, edetic acid disodium salt,
edathamil disodium, tetracemate disodium, Cheladrate, Chelaplex III,
Sequestrene NA, Sodium Versenate. And those are just a few of the ones
current in the USA; in Europe there are other names to add to the list.

The trivial names, which are often the older ones, tend to be those that
are generally used. Of course some of the much older names may be
found from time to time, for example, muriatic acid is the old name for
hydrochloric acid. That's not too much of a problem but the oxidation
product of this compound was called dephlogisticated muriatic acid gas;
its modern name is chlorine. Fortunately, there is international standardization in the form of the IUPAC (International Union of Pure and
Applied Chemists) system for assigning names to chemicals.

For cosmetic and toiletries the CTFA (Cosmetics Toiletries and Fragrance Association) and INCI (International Nomenclature of Cosmetic
Ingredients) names are used. Generally these names are pretty close to
the systematic ones, for example the CTFA/INCI name sodium laureth
sulfate suggests, to those with a little knowledge of surfactants, sodium
lauryl ether sulfate. However, this name falls short of uniquely identify-
ing the chemical in that the number of moles of ethylene oxide (see later)
is omitted.

If there are doubts over a name for a particular substance then the
only way to be sure is to use the Chemical Abstracts Service registry
number, CAS number, a unique identifier. For example, for EDTA
tetrasodium (anhydrous), the CAS number is 64-02-8.

WATER

In many preparations, particularly ones based on surfactants, water is the
major ingredient and, as such, justifies first place in the list of chemicals.

When studying formulations that are water based there is often a temptation to view the water's role as simply one of diluting the preparation – to make it go further. But this versatile liquid is not to be thought of lightly. Water plays a vital role in most everyday formulations and so an understanding of it is essential; and without this it is impossible to get a grasp of the way in which many other chemicals work.

In attempting to understand how water behaves we must, just as with other chemicals, consider the structure of its molecules. Two hydrogen atoms covalently bonded to a single oxygen atom give the water molecule its unique and fascinating properties, both physical and chemical. The hydrogen–oxygen covalent bond is highly polar because of the large difference in electronegativity of these two atoms.

Most of us have seen how a jet of water from a burette is bent by the electrostatic attraction of charged plastic rod placed near the jet. The polar bonds make water a polar liquid and this has important consequences as will be discussed in the section on solvents and dissolving.

At any one time a small number of these highly polar molecules will undergo complete polarization, resulting in a covalent bond breaking and dissociation into two ions. This gives rise to an equilibrium from which is obtained the ionic product of water, K_w,

$$H_2O \rightleftharpoons H^+ + OH^- \qquad (U1)$$

$$K_w = [H^+][OH^-]$$

where [ ] = concentration.

In any sample of pure water at room temperature the concentration of each of the ions is 10^{-7} mol l^{-1}. Thus the ionic product is $10^{-7} \times 10^{-7} = 10^{-14}$. An important aspect of this is that it is constant so that whatever is added to the water the ionic product stays the same in accordance with the equilibrium principle (Le Chatelier's Principle). An appreciation of this basic theory is prerequisite to understanding the behaviour in certain aspects of chemical formulation such as acids, bases, buffers and pH.

Another important property arising from the polarity of the water molecule is hydrogen bonding which explains much of how water interacts with other substances. Hydrogen bonding is considered in some detail in the section on dissolving and in the surfactants chapter (Chapter 2).

In some formulation chemicals water of crystallization is present due to the polar water molecules being strongly attracted to the anions and cations and becoming fixed in the lattice. A familiar example is washing soda, sodium carbonate decahydrate, $Na_2CO_3.10H_2O$, a crystalline solid that is mostly water (62.9%).

Apart from an understanding of some of the theoretical aspects of water, and how these are reflected in its properties, it is also important to know about the practical aspects that affect its use in formulations.

For formulation work the source of water is going to be the public supply provided by the water company. This water will be free from pathogenic micro-organisms but the water treatment processes from which it comes does not remove mineral matter such as dissolved calcium salts. Tap water is therefore a dilute solution of mineral salts. The amount of salts dissolved depends on the catchment area and how much limestone there is in that area. Water from a limestone area is high in calcium salts and is termed hard water whereas from non-limestone areas we get soft water. In general water is classified as:

Up to 100 mg l^{-1} $CaCO_3$ = soft water
100–400 mg l^{-1} calcium carbonate = medium hard water
400 upwards mg l^{-1} calcium carbonate = hard water
mg l^{-1} = milligrams per litre; it is numerically the same as parts per million (ppm)

For most chemical preparations tap water falls short in terms of chemical purity and some form of pre-treatment is called for to upgrade it. There are many methods for treating the water to remove the dissolved mineral salts but the most common is ion exchange which involves passing the water through a column of ion exchange resin.

Two types of ion exchange are common: water softening and water deionizing. In the former the calcium ions that cause hardness are exchanged for sodium ions that do not cause hardness. Although water softening plays a small role in preparing formulations it is more usual to opt for deionizing, sometimes called demineralization.

Modern ion exchange resins are based on synthetic polymers. In deionizing there are two types of resin, one for cation exchange and the other for anion exchange. The cation one is acidic and exchanges metal ions for hydrogen ions whereas the anion exchange polymer is basic and exchanges anions for hydroxide ions as shown in Figure 1.1.

These two resins, which are normally supplied as spheres about 1 mm in diameter, may be used in two different columns run in series or, as is more usual, as a 1:1 mixture in a single column. The former arrangement offers the possibility of recharging the resin once it is spent, thus scoring a few environmental points over the mixed resin approach where the whole lot is discarded.

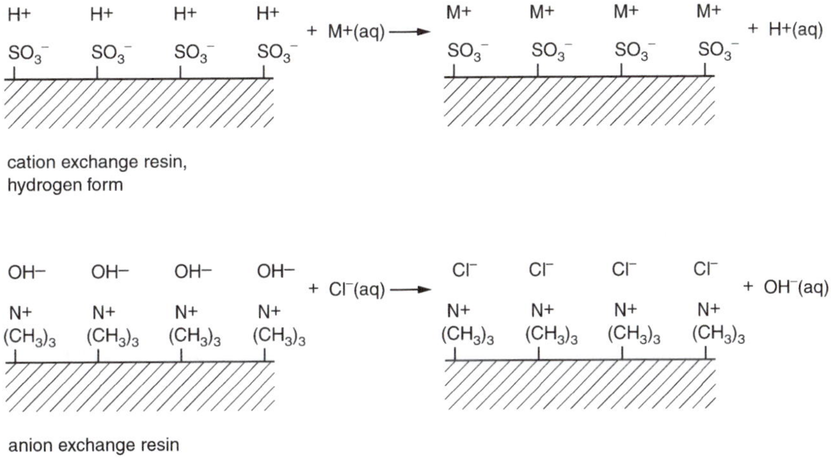

Figure 1.1 *Ion exchange reactions for producing deionized water*

Preparing personal care products imposes greater demands upon the quality of the water. Although the deionization process will mean that dissolved salts are absent it does not guarantee it is free from micro-organisms which can cause product spoilage or, even worse, carry harmful bacteria to the user.

Water from the public supply, because it carries a residual chlorine content, is disinfected when it is drawn from the tap but after passing through a bed of ion exchange resin its chlorine is removed and its disinfecting capacity lost. As such it may need a supplementary treatment to ensure destruction of micro-organisms.

One might argue the irrelevance of this on the grounds that a bactericidal preservative is added to the blend. But, the counter to this is that the preservative is put in to protect the formulation from micro-organisms picked up after its preparation and to give it a reasonable lifetime once its packaging is opened and it enters use. Biocidal preservatives are not intended as remedies to bacterial and fungal contamination from water or dirty mixing equipment.

SURFACTANTS

These are the main chemicals in the everyday formulations that form the basis of this book and, as such, are discussed at length in Chapter 2. Very briefly, surfactants are the most versatile of chemicals found in modern formulations and carry out functions such as: emulsification, detergency, wetting and spreading, solubilizing, foaming and defoaming, lubricity, biocidal, anti-static and corrosion inhibition.

For convenience, surfactants are grouped on the basis of their molecular types rather than functionality. The groups are: non-ionic, anionic, cationic and amphoteric. Within each category we may find that a range of functions is represented and that individual molecules can carry out a variety of tasks but with different degrees of effectiveness to related molecules.

ACIDS AND BASES (ALKALIS)

Strictly speaking an alkali is a base in aqueous solution and so the word is confined to soluble bases. As the word alkali is part of everyday language it is often used instead of the word base. Some texts talk in terms of acids and alkalis, others of acids and bases. The subtle difference between base and alkali need not be a problem in this text but both words must be used as the book sits somewhere between industrial chemistry and academic chemistry. As such, both words are used in a manner that reflects common usage rather than strict chemical definition.

It was seen above that water dissociates into hydrogen ions and hydroxide ions. In terms of simple definitions relating to aqueous chemistry: a substance that produces hydrogen ions as the only positive ions is an acid; a substance that reacts with hydrogen ions in a neutralization reaction is an alkali; usually alkalis react with hydrogen ions because they produce hydroxide ions.

Considering the dissociation of water we see that it behaves both as an acid and an alkali. It is amphoteric, a unique property that is also a feature of some surfactant molecules discussed later on in the surfactants section.

If we add an acid to water then extra hydrogen ions are introduced. For the ionic product to remain the same, as is required by the equilibrium principle, the number of hydroxide ions must diminish. Likewise, if a base is added to the pure water the hydroxide ions are increased and so the hydrogen ions must decrease.

At first sight it might seem that an impossible situation is going to arise in which there is an imbalance in the ratio of positive charge to negative charge. However, when, for instance, an acid is added this is not simply the addition of hydrogen ions but is also the addition of the accompanying anion. For example if hydrochloric acid, HCl, is added to the water the hydrogen ion concentration increases and the hydroxide concentration decreases by an amount that is offset by the chloride ions.

$$HCl\,(aq) \rightarrow H^+(aq) + Cl^-(aq) \qquad\qquad (U2)$$

By adding acid or alkali we can alter the concentrations of hydrogen ions and hydroxide ions giving rise to acidic or alkaline solutions of strengths appropriate to the amount added.

The pH scale, which is based upon the ionic product of water, is a useful measure of the strength of a solution's acidity or alkalinity.

pH = $-\log_{10}[H^+]$ where $[H^+]$ is hydrogen ion concentration in mol l^{-1}

pH	0	1	2	3	4	5	6	7	8	9	10	11	12	13	14
	strong acid		**weak acid**					**neutral**			**weak alkali**			**strong alkali**	

The scale, being logarithmic, means that one pH unit relates to a factor of ten. Thus, a solution of pH 3 is ten times more acidic than a solution of pH 4 and a solution of pH 12 is ten thousand times more alkaline than a solution of pH 8.

The pH scale is useful for quantifying the acid or base strengths of dilute solutions; it is not appropriate for strong solutions and if applied in such situations misleading results are obtained. In most situations the solution under study can be diluted to bring it to the appropriate concentration. In practice where the acidic or alkaline strength of a solution is reported it is given with the dilution figure.

Thus an alkaline degreaser solution may state: pH (0.1%) = 12.5. Measurement of pH is normally carried out by means of a pH meter and combination electrode. This electrode has a thin glass membrane and must never be used with hydrofluoric acid. Indicators, as solutions or papers, are useful as an approximation where the solution being measured is colourless.

The strengths of acids and bases themselves are also important. Strong acids are highly dissociated and produce many hydrogen ions whereas weak acids only partially dissociate and provide small numbers of hydrogen ions. This can be shown as an equilibrium from which we get the acid dissociation constant, K_a. Values for K_a of different acids are to be found in tables in reference texts.

$$HA(aq) \rightarrow H^+(aq) + A^-(aq) \qquad \text{(U3)}$$

From this we have

$$K_a = \frac{[H^+][A^-]}{[HA]}$$

where [ ] = concentration

Care needs to be taken not to confuse strengths of acids with pH. For example a solution of the weak acid, acetic acid, may well have a pH less than (more acidic) a more dilute solution of a strong acid such as hydrochloric. A similar system, base dissociation constant, K_b, applies to bases but finds less application than the one for acids.

This helps to appreciate the distinction between the strengths of acids and the strengths of acid solutions and is required if work on buffer systems (see below) is to be carried out.

Strong mineral acids feature in aqueous preparations where highly acidic conditions are required such as in metal treatments preparations and produce acidity as shown below.

$$\begin{aligned}
&HCl(aq) \rightarrow H^+(aq) + Cl^-(aq) &&\text{single dissociation} \\
&H_2SO_4(aq) \rightarrow H^+(aq) + HSO_4^-(aq) &&\text{first dissociation} \\
&HSO_4^-(aq) \rightarrow H^+(aq) + SO_4^-(aq) &&\text{second dissociation}
\end{aligned} \quad (U4)$$

There is a decrease in the value of K_a from the first dissociation through successive ones. Acids are sometimes referred to on the basis of the number of moles of hydrogen ions that can be formed from one mole of acid: hydrochloric acid is monobasic, sulfuric acid is a dibasic acid, phosphoric acid and citric acid (2-hydroxypropane-1,2,3-tricarboxylic acid) are tribasic acids. It is of interest to compare the actual values of K_a (25 °C) for phosphoric acid (a mineral acid) and citric acid (an organic acid).

$$\begin{aligned}
&H_3PO_4(aq) \rightarrow H^+(aq) + H_2PO_4^-(aq) &&K_a \text{ (first)} = 7.52 \times 10^{-3} \\
&H_2PO_4^-(aq) \rightarrow H^+(aq) + HPO_4^{2-}(aq) &&K_a \text{ (second)} = 6.23 \times 10^{-8} \\
&HPO_4^{2-}(aq) \rightarrow H^+(aq) + PO_4^{3-}(aq) &&K_a \text{ (third)} = 2.2 \times 10^{-13}
\end{aligned} \quad (U5)$$

$$(C_3H_6O)(COOH)_3(aq) \rightarrow H^+(aq) + (C_3H_6O)(COOH)_2COO^-(aq)$$
$$K_a \text{ (first)} = 7.45 \times 10^{-4}$$
$$(C_3H_6O)(COOH)_2COO^-(aq) \rightarrow H^+(aq) + (C_3H_6O)COOH(COO^-)_2(aq) \quad (U6)$$
$$K_a \text{ (second)} = 1.73 \times 10^{-5}$$
$$(C_3H_6O)COOH(COO^-)_2(aq) \rightarrow H^+(aq) + (C_3H_6O)(COO^-)_3(aq)$$
$$K_a \text{ (third)} = 4.02 \times 10^{-7}$$

Clearly most of the acidity comes from the first dissociation in both cases. It is this acidity that is most important in assessing acids for applications such as formulated products. Citric acid is the weaker of these two examples and, in general, organic acids are weaker than mineral acids.

Acidification of preparations, especially those for personal care use, is therefore frequently brought about by the weaker organic acids: citric or lactic. Oxidizing acids, for instance nitric acid, are never used in combination with organic substances because of the potentially explosive combinations that may result. Hydrofluoric acid is highly toxic and rarely used apart from in glass etching and some metal pickling formulations.

Certain salts, the acid salts, such as sodium hydrogen sulfate (sodium bisulfate) may also be used for acidification. This particular salt has the advantage of being a strong acid in solid form, its acidity coming from dissociation of hydrogen sulfate ion upon contact with water.

$$NaHSO_4(aq) \rightarrow Na^+(aq) + HSO_4^-(aq)$$
$$HSO_4^-(aq) \quad \rightarrow H^+(aq) \; + SO_4^{2-}(aq) \tag{U7}$$

It will be seen that this is the second dissociation of sulfuric acid. As such the hydrogen sulfate ion is weaker than sulfuric acid.

Some acids, the fatty acids or long chain carboxylic acids, are used for their soap forming ability rather than for acidification. For example, stearic acid may be put into a preparation and then an exact amount of sodium hydroxide added to convert the acid into the soap, sodium stearate. Low values for K_a mean that the long chain fatty molecules are feeble acids, unsuitable for pH adjustment or general acidification.

$$CH_3(CH_2)_{16}COOH(s) + NaOH(aq) \rightarrow CH_3(CH_2)_{16}COONa(aq) \tag{U8}$$

This reaction is in essence that of salt formation and similar to those discussed in the salts section. A blend of fatty acids known as tall oil fatty acids (TOFA), a by-product of the wood pulping industry, is available in different grades depending upon the amounts of rosin in it. This blend represents a reasonably low cost source of fatty acids but is generally confined to the preparation of cleaning formulations.

Alkalinity is an important property of many surfactant formulations as it plays a big part in the cleaning action. This is particularly so in the case of heavy duty degreasing preparations. Many of these are confined to industry and to mechanized degreasing processes where there is no contact between the chemical and personnel. High alkalinity preparations are caustic and dangerous in the event of eye or skin contact.

Many types of soiling are acidic so that when they come into contact with an alkaline solution they form soluble salts that are then dissolved from the substrate surface into the water. Oily soiling frequently contains dusts bound to the surface by the oily substances so that once the oil is made soluble and released the dust is simultaneously released and dispersed into the washing water.

Where the oily material contains long-chain fatty acids and their glyceryl esters reaction with alkali such as sodium hydroxide turns them into soaps; not only is the dirt removed but it is turned into a surfactant itself. The slimy feel we experience when some dilute sodium hydroxide solution gets onto our hands is in fact this very reaction in which fatty acids from the skin form a soap, the slimy feel, with the alkali. So effective are alkalis at dirt removal that a simple solution of say sodium carbonate, washing soda, is effective even in the absence of surfactants.

The highest levels of alkalinity are brought about by small additions of sodium or potassium hydroxide to an aqueous preparation. Although the sodium compound is more widely used on account of its low cost, both have the ability to break down fats and proteins by hydrolysis, even at concentrations below 1%. Potassium hydroxide is used on account of the high solubility of both itself and the products it forms during the cleaning process.

Where a preparation is required to have easy rinse away properties the potassium compound is favoured as it is less likely to produce streaking than the sodium one. Sodium metasilicate and, less so, sodium carbonate are also capable of producing highly alkaline blends required by medium to heavy duty cleaning preparations.

Moderate to lower levels of alkalinity come from adding salts such as trisodium phosphate, tetrapotassium pyrophosphate and sodium tripolyphosphate. The phosphates, being less alkaline find use in preparations where skin contact is likely. Sodium carbonate is cheap, moderately alkaline and available as decahydrate (washing soda) or anhydrous (soda ash). Ammonia solutions also produce moderate alkalinity but this is soon lost as ammonia gas diffuses out into the atmosphere. However, derivatives of ammonia such as monoethanolamine find wide application in formulation work.

The level of alkalinity of a preparation may be crucial to its working. This is particularly the case where alkaline metal cleaning formulations are concerned. For example metals such as aluminium and zinc, including galvanized steel, may be etched – or in extreme cases completely dissolved – by highly alkaline solutions. For any metal cleaner the pH needs some serious thought.

SALTS

There are various salts of importance in formulations, especially in detergent formulations, and a basic knowledge of their properties and behaviour is essential to understanding the chemistry behind many preparations. Quite often salts, alkalis and chelating agents (sequestrants) are referred to in the detergent industry as builders.

In general we can regard a salt as comprising an acid part (anion) and a part derived from a base (cation). Usually the cation is a metal ion but sometimes it is ammonium and its derivatives. As seen above there are different strengths of acids and bases as shown by the dissociation constants and the neutralization reactions that they can take part in lead to many different salts each one of which has properties relating to the strength of the acid and base from which it was derived.

Common salt, sodium chloride, is a salt derived from a strong acid, hydrochloric acid, and a strong base, sodium hydroxide.

$$HCl + NaOH \rightarrow NaCl + H_2O \qquad \text{(U9)}$$

Salts formed by neutralization of strong acids and strong bases are neutral salts and in solution give solutions of around pH 7. On the other hand if either the acid or the base is strong and the other is weak the salt produced is far from neutral.

Sodium carbonate, for instance, is derived from a weak acid, carbonic acid, and a strong base, sodium hydroxide. When dissolved it gives a solution of high pH, *i.e.*, it is alkaline.

$$H_2CO_3 + 2NaOH \rightarrow Na_2CO_3 + H_2O \qquad \text{(U10)}$$

This reaction is seen where sodium hydroxide is left open to the air and turns into a crumbly white powder of sodium carbonate. As a consequence there can be considerable wastage where sodium hydroxide and its preparations are not kept sealed.

Magnesium chloride, an example of a salt that is acidic in solution, is from reaction between the strong acid, hydrochloric acid, and the weak base, magnesium hydroxide. When dissolved this one gives solutions of low pH, *i.e.*, it is acidic.

$$Mg(OH)_2 + 2HCl \rightarrow MgCl_2 + H_2O \qquad \text{(U11)}$$

A fuller understanding of salt behaviour requires a study of how salts hydrolyse in terms of the dissociation constants briefly referred to earlier. Usually it is sufficient to apply the general rule: salts of strong acids and weak bases are acidic; salts of weak acids and strong bases are alkaline.

Some care is needed here as there is often confusion over salts that are acidic and ones that are described as acid salts. An acid salt is a salt produced by only partial neutralization of the acid, a dibasic or tribasic acid, by the base. For example, reacting sulfuric acid and sodium hydroxide to the half neutral point gives sodium hydrogen sulfate, an acid salt (see also acids section above).

$$H_2SO_4 + NaOH \rightarrow NaHSO_4 + H_2O \qquad (U12)$$

To continue the reaction to complete neutrality produces the normal salt which is neutral.

$$NaHSO_4 + NaOH \rightarrow Na_2SO_4 + H_2O \qquad (U13)$$

Sodium bicarbonate (sodium hydrogen carbonate), $NaHCO_3$, is an acid salt represented by half neutralized carbonic acid but it is not acidic, it is alkaline.

In addition to these inorganic salts there are also those derived from organic acids and organic bases (Table 1.1). Organic acids are invariably weak acids but the organic bases can be quite strong. Important examples, from the formulation standpoint, are long-chain fatty acid salts, soaps. Some of these fatty acid salts are formed *in situ* as the preparation is being put together such as the neutralization between oleic acid and monoethanolamine to produce the soap, monoethanolamine oleate. This is a salt in which the anion is oleate and the cation monoethanolamine; it is therefore an organic salt.

Other salts commonly used in cleaning formulations are sodium phosphates, as water softeners and sodium silicates. The latter prevents corrosion in detergent powder formulations based on sulfonate surfactants which are aggressive towards metals such as aluminium. Both these also provide alkalinity and some degree of pH buffering (see below) against soil acidity as they are salts of weak acids and strong bases. There are a variety of sodium silicates, orthosilicate and metasilicate for example, with differing strengths of alkalinity depending upon the relative proportion of silicon oxide and sodium oxide.

Of the phosphates sodium tripolyphosphate became the favoured one for detergents as it also helped in keeping removed soil in suspension. With regard to soil suspension in detergent formulations the compound sodium carboxymethylcellulose (SCMC) is particularly effective.

Salts are sometimes used not for their chemical properties but for their physical properties. Thus, anhydrous sodium sulfate is a significant part of many powder detergent formulations and acts as a filler ensuring good handling characteristics of the powder.

BUFFERS

In aqueous formulations there is often a requirement for a fixed pH that will remain stable and resist the effect of small amounts of acid or alkali getting into the preparation or being formed as the ingredients age. In other words, they need some buffer capacity.

Table 1.1 *pH of 1% aqueous solutions of acids, alkalis and salts commonly used in formulations*

System	pH
Sodium hydroxide, caustic soda, NaOH	13.4
Potassium hydroxide, caustic potash, KOH	13.0
Sodium metasilicate, Na_2SiO_3	12.8
Trisodium phosphate, Na_3PO_4	12.0
Tripotassium phosphate, K_3PO_4	11.8
Monoethanolamine, MEA	11.6
Sodium carbonate (anhydrous), soda ash, Na_2CO_3	11.2
Tetrapotassium pyrophosphate, TKPP, $K_4P_2O_7$	10.3
Triethanolamine, TEA	10.2
Sodium tripolyphosphate, STPP, $Na_5P_3O_{10}$	9.7
Potassium tripolyphosphate, $K_5P_3O_{10}$	9.7
Sodium tetraborate, $Na_2B_4O_7$	9.4
Sodium bicarbonate, sodium hydrogen carbonate, $NaHCO_3$	8.4
Monosodium phosphate, NaH_2PO_4	4.6
Acetic acid, ethanoic acid, CH_3COOH	2.4
Citric acid, 2-hydroxypropane-1,2,3-tricarboxylic acid	1.9
Phosphoric acid, orthophosphoric acid, H_3PO_4	1.3

Many salts, usually in combination with the weak acids or weak alkalis of that particular salt, will provide buffer capacity. For example, a dilute solution of sodium acetate and acetic acid behaves as an acid buffer (pH 4.5) due to the following reaction.

$$CH_3COOH(aq) \rightarrow CH_3COO^-(aq) + H^+(aq) \qquad \text{(U14)}$$

This dissociation of acetic acid occurs only to a small extent because of the small dissociation constant. Thus, most of the acid remains as molecules. However, if some of the hydrogen ions were removed by reaction with alkali this disturbs the equilibrium which then re-establishes itself by more of the acid molecules dissociating. As such it has the capacity to resist pH change due to small additions of alkali.

If, instead of alkali being added, there is a small amount of acid put in this increases the concentration of hydrogen ions but equilibrium is re-established by the dissociation reaction going into reverse and consuming those extra hydrogen ions to form more acetic acid molecules.

Again, the acetic acid dissociation reaction has compensated to keep the hydrogen ion concentration and hence the pH constant. The dissociation swings either way to add or remove hydrogen ions and buffer the effect of small disturbances. To ensure that there is always a sufficiency of acetate ions sodium acetate is used.

$$CH_3COONa(aq) \rightarrow CH_3COO^-(aq) + Na^+(aq) \qquad (U15)$$

In a similar manner buffers can be made for maintaining just about any pH. A buffer working in the alkali region would result from a weak base and a salt of that weak base. All buffers have their limits and cannot go on indefinitely removing added acid or alkali.

The sodium silicate, a salt of a strong base and a weak acid, put into some detergents acts as a buffer to absorb the effect of soil acidity. Other examples of chemicals that provide buffer action are: monosodium phosphate (MSP), disodium phosphate (DSP), citric acid/sodium citrate.

CHELATING AGENTS, SEQUESTRANTS

Chelates are complexes in which an organic molecule forms co-ordinate bonds with certain metal ions. Electron pairs, normally from nitrogen and oxygen atoms, bond with empty orbitals of the metal cation. Nitrogen and oxygen atoms are particularly effective as a source of lone pair electrons and feature in many chelating agents. Nature has many examples of chelates, one of the more familiar ones is the green compound in which a magnesium cation is chelated to form chlorophyll as shown in Figure 1.2.

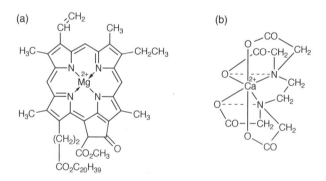

Figure 1.2 *Chelated metal ions: (a) chlorophyll, one of nature's chelates, (b) EDTA, a widely used synthetic chelating agent, is effective in sequestering calcium ions*

In chemical formulation work it is the synthetic chelating agents that are important – although some natural based ones are used and now in increasing amounts. Of the man-made ones it is the aminocarboxylic acids that predominate.

In a solution of EDTA (ethylenediaminetetracetate) ion, a common synthetic chelating agent, six co-ordinate bonds are formed to a calcium ion. Here the lone pairs are from the two nitrogen atoms and four of the oxygen atoms. A stable hexadentate (six chelating bonds) complex is formed and the metal ion is thus firmly grasped, pincer style (Greek: chela = crabs claw). As such the metal ion is no longer free to take part in other reactions; it has been sequestered.

Many other metals can be chelated by EDTA but not always in a hexadentate arrangement. Usually one mole of EDTA will complex with one mole of metal ion. From this it is a straight forward task to calculate the amount of chelating agent (sequestrant) to use in a preparation or process.

Chelating agents are often effective over only a narrow pH band; outside this their effect is weakened or they are rendered totally ineffective. Their selectivity for different metals is also a function of pH. For example, EDTA complexes most effectively with calcium ions in fairly alkaline conditions at pH 10, with lead at pH 6, but for it to be effective with iron(III)ions an acidic solution is required at pH 3. Other chelating agents may well be quite different in terms of pH and performance. The effectiveness of chelating agents relates to the stability constant which is an equilibrium constant.

$$M(aq) + Z(aq) \rightleftharpoons MZ(aq)$$

where M = metal ion, Z = chelating agent

$$K = \frac{[MZ]}{[M][Z]} \qquad \text{(U16)}$$

Chelating agents can thus be used to prevent metal ions being free to interfere with other substances and have an adverse effect. In virtually all processes that involve water there are potential problems from dissolved metal ions. Even where a formulation is made with deionized water metal ions can still become a problem as they can be picked up from the fabric of the mixing plant or from the chemicals used. Furthermore some formulations, particularly washing preparations, are diluted with tap water by the user and this can introduce quite large concentrations of dissolved calcium salts.

Metal ions present in a formulation may result in discolouration, as in the case of iron or manganese, precipitate out of solution as insoluble compounds or bring about unwanted reactions that interfere with, say, the cleaning performance. Addition of a chelating agent to cleaning formulations prevents interference by metal ions picked up during dilution and in the cleaning process itself.

In personal care formulations chelating agents are added to sequester traces of metal ions such as iron that would otherwise catalyse the degradation of the fatty components and turn the preparation rancid or discoloured. A particular example of this is oxidation of the double bond in oleic acid.

$$CH_3(CH_2)_7CH=CH(CH_2)_7COOH + O_2 \rightarrow CH_3(CH_2)_7CH{-}CH(CH_2)_7COOH$$
$$\underset{O{-}O}{\overset{\mid\quad\mid}{}} \qquad (U17)$$

Further oxidation then occurs to form a variety of rancid smelling compounds such as pelargonic acid. A face cream that smells of sweaty socks or a bar of soap that has the odour of vomit is not going to be a market leader. Of course one way round these problems is to have perfectly pure materials containing none of the catalytic metal ions: pure water, pure chemicals, pure container materials and non-metallic mixing equipment. These conditions are, however, not approachable unless ridiculous levels of purity are employed.

Today there are many synthetic chelating agents commercially available for adding to preparations to prevent problems from the dissolved metal ions from various sources. Most chelating agents are synthetic but a few are derived from natural chemicals.

The most widely used sequestrant in surfactant formulations is currently nitrilotriacetate (NTA). Other common ones are: diethylenetriaminepentaacetate (DTPA) and hydroxyethylethylenediaminetriacetate (HEDTA). Although supplied as their sodium salts other salts are available such as: ammonium, calcium, iron as well as the acids themselves.

Sodium gluconate, derived from gluconic acid, is a popular chelating agent and has the virtue of being made from a natural renewable chemical and is completely biodegradable. Relatively high concentrations of chelating agents are used in cleaning formulations whereas for cosmetics and toiletries only small amounts are needed.

In many household detergents NTA has replaced phosphates because of the problems generated when these phosphates pass through sewage works and into watercourses. The main problem is that phosphate is a plant nutrient and excessive growth of aquatic plants occurs and as dead plant material builds up the process of eutrophication sets in.

However, chelating agents such as EDTA and NTA are not without their problems when it comes to environmental worries and much research is currently devoted to this area. From the early results of this it appears that the NTA has less environmental impact than EDTA.

SOLVENTS

Water is the major solvent in many preparations and, especially in combination with surfactants, is capable of bringing into solution a large number of other substances but it has its limitations and when these are reached the range of organic solvents must be considered. By such means the scope of surfactant formulations is opened up even further.

Because of the large number of organic solvents only the common ones will be considered here. Today's range of solvents available for formulation work is represented in the main by ones that are obtained directly from petroleum or synthesized from petroleum feedstocks. There are also some solvents made from plant materials and, although the proportion of these is relatively small compared with the petroleum-derived solvents it is growing as a result of our desire to become less reliant upon petroleum.

Solvents from plant sources compare well with the petroleum-based ones but there are some noticeable absences, the main ones being the halogenated hydrocarbons, particularly the chlorinated ones. These chlorohydrocarbons are not found in nature to any significant extent. In fact when they do get into natural systems they can wreak havoc but they are excellent solvents and some industrial applications are going to be hard pressed as these solvents are forced off the scene due to environmental concerns.

Many plant-derived solvents cost more than their man-made counterparts and supply of them often fluctuates as they are prone to the uncertainties that are forever present in the business of growing plants. Despite some of the drawbacks to using plant-derived solvents there is considerable enthusiasm for them because of their perceived 'cradle-to-grave' profile and this should ensure a steady expansion of them in formulation work. However, perceptions are one thing and reality another. Some chemicals that appear virtuous from simplistic notions of the 'cradle' can turn up a few surprises when a thorough analysis is carried out. This is discussed later in the chapter on environment and resources (Chapter 5).

Many organic solvents have a long history and provide some interesting examples of the bygone technology that was used in separating them from the plant or animal material in which they originated. Methanol (methyl alcohol, wood alcohol), a poisonous alcohol, was obtained by destructive distillation of wood. Related to methanol is methanoic acid (formic acid) that was also obtained by distillation, but not from material of plant origin. For this acid ants were put into the distillation pot.

Perhaps the solvent that has attracted greatest attention is ethanol (ethyl alcohol), a chemical that we in the wealthy industrialized parts of the world drink in ever increasing amounts. Made traditionally by the fermentation of sugars, ethanol is also nowadays man-made, for industrial use, from ethene (ethylene) and steam.

There is no doubt that people's desire for intoxication provided much of the momentum for the development of fermentation and distillation technology. Ethanol, apart from its use as psychoactive beverage, is widely used as a solvent and, in increasing volumes, as a motor fuel in place of petroleum fuels.

THE PROCESS OF DISSOLVING

For any meaningful discussion of solvents we need to know how they work. The process of dissolving involves the interaction of solvent and solute to form a solution. For example water, a solvent, will dissolve salt, a solute, to form salt solution whereas mixing paraffin wax into water will never produce a solution no matter how much effort is invested.

Without some basic knowledge, anyone involved in making solvent-based preparations, or even adding solvents to aqueous systems, is left with nothing more to go on than a set of trial and error experiments – a costly, lengthy and possibly risky exercise. In considering the process of dissolving we need to start at the molecular level, work through some basic theory and examine how one substance interacts with another. And to do that requires a detailed look at the particles that make up the solvent and solute: molecules, ions, atoms.

Consider a substance made up of particles A that is to be mixed with a substance comprising particles B. In substance A there are attractive forces between the particles that may be represented as A–A; in a similar manner substance B may be represented as B–B.

For dissolving to occur these attractive forces must be overcome by new attractive forces between A and B represented as A–B. If the A–B forces are stronger than the A–A and B–B forces then dissolving occurs. Conversely if the A–B forces are less than the A–A and B–B forces then dissolving cannot result.

To fully appreciate the dissolving processes we need to be aware that there are two quite separate controlling factors: energetics and kinetics. The energetics relate to the attractive forces just mentioned and indicates the possibility of a process occurring whereas the kinetic factors dictate how fast the process will take place.

The simple equation relating to the energetics is:

$$\Delta G = \Delta H - T\Delta S$$

where ΔG = Gibbs free energy change, ΔH = enthalpy (heat) change, T = Kelvin temperature and ΔS = entropy change.

Dissolving is possible only when ΔG is negative, *i.e.*, where there is a decrease in free energy and this usually happens because ΔH is large and negative, the process is exothermic, whilst $T\Delta S$ is small. As a general rule when dissolving occurs there is an accompanying heat loss from the substances involved. This is readily demonstrated by the dissolving of glycerol (propane-1,2,3-triol) in water, when a noticeable amount of heat is given out. However, general rules have their exceptions.

If ammonium nitrate is stirred into a beaker of water it dissolves easily but the mixture gets ice cold and condensation freezes on the outer surface of the beaker. The process is endothermic, ΔH is large and positive, taking in heat from its surroundings. This does not in any way break the rule that ΔG must be negative for a process to occur. ΔH is positive because this example is one of the few which has a high value for ΔS. We would say it is an entropy driven process rather than an enthalpy driven one.

In an energetically favourable situation the dissolving process may proceed rapidly, slowly or not at all depending upon the kinetics. Temperature normally increases the rate of dissolving and also affects the extent, but the converse can also apply as some solutes become less soluble at elevated temperatures. This is an important aspect of certain surfactant molecules, the ethoxylates, and is discussed in the surfactants section.

The effect of temperature on increasing the rate of dissolving is explained in terms of the energy that the molecules have. Consider a sodium chloride crystal in water and visualize the water molecules arriving at the surface of the crystal. With cold water the molecules move slowly; for hot water the opposite applies. The faster the particles move the faster they arrive at the surface of the crystal, hence the greater the rate of dissolving.

Where the solute is a solid the particle size may have a marked effect on the rate at which it dissolves; the smaller the particle size, the greater the surface area and the greater the speed of dissolving. Agitation is often a requirement to hasten the process of dissolving. Again consider a crystal of sodium chloride dissolving in water. Near to the surface of the crystal the water will soon become saturated with salt and further dissolving is restricted – slow diffusion across the concentration gradient will enable a minor amount of dissolving to continue.

With agitation, the saturated solution is dispersed and pure water arrives at the crystal surface to dissolve more salt. Thus, agitation is essential when dissolving solids in a solvent for the process to go at a reasonable speed. Clearly in the dissolving mechanism kinetic factors can play a big part but no amount of manipulating of these will bring about dissolving if ΔG is positive.

To learn more of the factors that result in energy changes during dissolving requires an understanding of the attractive forces between the molecules. The types of attractive forces that are of importance here are, going from the strongest to the weakest, ionic forces, hydrogen bonds, permanent dipole attractions and Van der Waals forces. Each one of these is a result of electrostatic attractions which in turn are due to permanent polarization of the electrons that results from differences in electronegativity between atoms, or temporary polarization induced by the proximity of neighbouring molecules.

Ionic Bonds

Attractive forces between ions of opposite charge are what would generally be recognized as ionic bonds. This is polarization at the extreme as it involves anions and cations, the result of complete electron transfer from one atom to another. Ionic bonds are the strongest of all chemical bonds and are the basis of ionic lattices composed of alternate anions and cations held firmly in position by the attraction of opposite charges. Despite the fact that ionic bonds are very strong they can often be easily broken during the process of dissolving.

Covalent Bonds

These bonds, which involve the sharing of electrons between atoms but not complete transfer, are only ever broken, in the process of dissolving, where strong polarization occurs (see below) resulting in some ionic character.

Hydrogen Bonds

Wherever a hydrogen atom is covalently bonded to a strongly electronegative atom such as oxygen, nitrogen or fluorine the result is a highly polarized bond with the electrons pulled towards the electronegative atom and away from the hydrogen. This produces an intense positive charge around each hydrogen atom which then experiences dipole–dipole attractions with the oxygen atoms on adjacent molecules as shown in Figure 1.3.

Hydrogen bonds are much weaker than covalent or ionic bonds but can still have a profound effect upon the properties of certain substances.

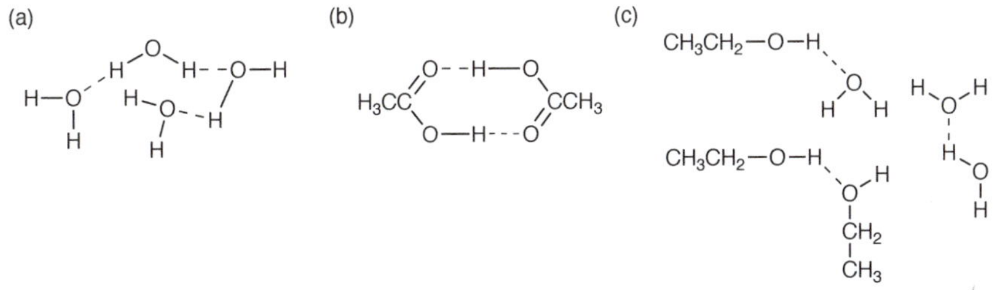

(a)   (b)   (c)

Figure 1.3 *Hydrogen bonds formed between different molecules occur when hydrogen is covalently bonded to highly electronegative atoms. (a) Hydrogen bonds cause water molecules to associate, (b) carboxylic acid molecules can form dimers due to hydrogen bonding, (c) alcohol dissolves in water due to hydrogen bonding*

Water, for example, has a much higher boiling point than would be predicted on the basis of molecule size and this is due to intermolecular hydrogen bonding and the extra thermal energy required to move a molecule from the liquid phase to the gaseous phase.

Dipole–Dipole Attractions

Other molecules in which there is a significant difference in electronegativity between the atoms in a covalent bond have dipoles and can thus be attracted to similar neighbouring molecules. With trichloromethane the electrons are pulled towards the three chlorines, creating a permanent dipole as shown in Figure 1.4. As a result there are dipole–dipole attractions associated with trichloromethane molecules.

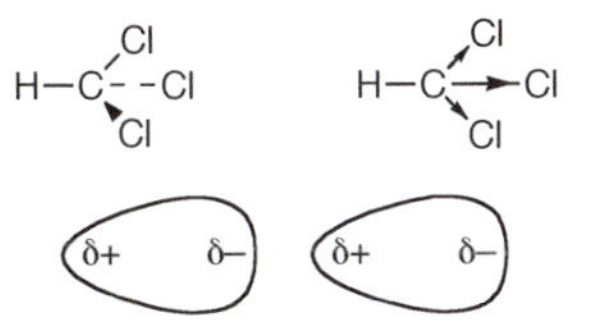

Figure 1.4 *Molecules such as trichloromethane have their electrons pulled towards the more electronegative (chlorine) atoms which sets up a dipole resulting in one molecule attracting another*

Van der Waals Attractions

These are a result of molecules having a temporary polarization due to small electron shifts. As one molecule becomes polarized it can induce a dipole in an adjacent molecule which can then be carried to other molecules. These induced dipoles cause intermolecular forces of attraction.

With small molecules the Van der Waals forces are very weak but as the size of the molecule increases these attractive forces become much more significant. In a liquid such as water the main intermolecular attractions are hydrogen bonds but with, say, hydrocarbons such as hexane the intermolecular attractions are Van der Waals forces.

Example 1, Dissolving Sodium Chloride in Water

When a water molecule arrives at the surface of the ionic sodium chloride crystal there is an attraction between the polar water molecules and the ions at the crystal surface. The strength of this attraction is sufficient to overcome the ionic forces that hold the ions in the lattice. The result is that the cations and anions acquire a sphere of water molecules around them; they become hydrated and leave the lattice and we see the crystal dissolve.

Example 2, Dissolving Paraffin Wax in Hexane

The molecules in paraffin wax are non-polar and are held in place as the solid by means of weak Van der Waals forces. Hexane, a liquid hydrocarbon, is also non-polar with Van der Waals attractions between the molecules. When this liquid comes into contact with the wax surface there is an attraction between the molecules in the solid and the hexane molecules; this attraction is slightly stronger than the attractions existing in the pure substances and so the wax dissolves into the hexane.

Polyethylene (Polythene), a hydrocarbon polymer, is chemically similar to paraffin wax, differing only in the lengths of its molecules, but it will not dissolve in hexane as does the wax. This illustrates an exception to the rule and demonstrates the importance of molecular size in the process of dissolving.

The fact that paraffin wax does not dissolve in water should not be the basis for a general conclusion about wax solubility. Some waxes, for example the polyethylene glycol waxes (Carbowax), appear to be quite like paraffin wax in some properties but they are completely soluble in water.

Example 3, Dissolving Sucrose in Water

A crystal of sucrose (sugar) is composed of molecules that each have several hydroxy groups which provide the sucrose molecule with a good measure hydrogen bonding capacity. When a water molecule, which is also polar and capable of hydrogen bonding, comes in contact with the sucrose crystal an attraction occurs. The strength of the attraction between a sucrose molecule and a water molecule is greater than the attractions in the individual molecules in the two substances and so a solution forms.

Example 4, Dissolving Ethanol in Water

Ethanol is a liquid composed of small molecules each containing a single hydroxyl group. As with the sucrose example above this results in the ability to form hydrogen bonds. When water molecules, which have a similar capacity, mix with the ethanol ones there is a stronger attraction between the ethanol and water than in the pure liquids. The result is that a solution forms. Strictly speaking we should say that ethanol is miscible with water rather than using the word soluble.

Example 5, Hydrocarbon Oil and Water

Hydrocarbon oils do not dissolve in water. On examining the behaviour of the two substances in contact with each other it is tempting to conclude that there is actually a repulsion between them, as if the like-poles of a magnet had been brought together. There is no such force of repulsion in operation here; it is simply that there is very little attraction between the two.

The oil molecules are attracted to other oil molecules and the water molecules are also attracted to their own kind. If some oil molecules find themselves surrounded by water molecules they simply get squeezed out and *vice versa*.

General Rule

A useful rule for selecting a solvent for a particular substance is 'like dissolves like'. As with all rules it has its limitations and exceptions. It is useful for guidance so long as it is borne in mind that the likenesses sought are in terms of molecular structure with emphasis on molecule polarity. Non-polar molecules of one substance are likely to attract non-polar molecules of a second substance. Similarly, polar molecules attract other polar molecules. Thus, we can generally regard solvents as polar, non-polar and somewhere in between.

Most solutes have limited solubility in a particular solvent, for example salt dissolving in water, but for some it is infinite as in the mixing of ethanol and water. There is sometimes no clear distinction between which substance is the solute and which is the solvent especially when the ratio of solvent to solute approaches parity.

The following example, which is not of much interest in formulation work, illustrates the point. If we wanted to dissolve gold the starting point in looking for a solvent would be to consider a liquid metal – like dissolves like. In fact mercury, a liquid metal at room temperature, readily dissolves gold.

With the polythene example referred to above it was seen that molecular size has a bearing on solubility. This is further exemplified in the use of a paint stripper. Here we see that dichloromethane is the most successful of the chlorinated hydrocarbons and much of its success is due to its molecular size. Similar, but larger, molecules do the same job but it takes them longer. This is because the mechanism involves the solvent molecules forcing their way into the paint matrix – the smaller ones can squeeze in easily and get to work.

In many applications a solvent may be chosen to simply act as a vehicle for the solute. For example, in a lacquer made up of shellac wax dissolved in ethanol, the ethanol is the solvent/vehicle and enables the wax to be maintained in solution up to its application to the surface, as in French polishing. After application the solvent evaporates and leaves behind a thin film of the shellac.

On the other hand the solvent may play a more active role as in a paint stripper made up of dichloromethane made into a viscous paste by means of a thickener. Here the solvent, dichloromethane, is the active part; it is the part that causes the paint to swell, soften and be released from its substrate.

Incorporating a solvent into a preparation may not always be the simple task. For example where a solvent such as white spirit, insoluble in aqueous solutions, is required as a part of a water-based preparation then it must be emulsified or solubilized by means of surfactants – see surfactants section.

Hydrocarbons

The bulk of hydrocarbons is from crude petroleum and the biggest proportion of them is for use as fuels: petrol, Diesel, kerosene *etc*. All hydrocarbons, compounds of hydrogen and carbon only, will burn to carbon dioxide and water where there is sufficient oxygen. The ones with small molecules, the lower members (gases and low boiling liquids) have low flash points requiring them to be classified as extremely or highly flammable whereas the larger molecules (liquids and solids) give rise to higher flash points of the flammable and combustible classifications.

These lower hydrocarbons do not play any significant role in formulation work apart from being used as propellants in aerosol preparations. Hydrocarbons represent some of the lowest cost solvents which accounts for their extensive use in paint formulations and in degreasing processes. In this group fall some of the earliest solvents to be used on a large scale, ones such as turpentine that are obtained from natural renewable resources.

Aliphatic Hydrocarbons

This group comprises molecules in which the carbon atoms are joined in straight chains, branched chains, rings and combinations of these. Within the group there are unsaturated hydrocarbons, alkenes (olefins) that have less than their full complement of hydrogen atoms and the saturated hydrocarbons alkanes (paraffins). Chain branching and ring formation add further to the number of hydrocarbons in this group.

Aromatic Hydrocarbons

The aromatics differ from the aliphatic hydrocarbons in that they contain molecules based upon the benzene ring. In many aromatics the molecules also contain aliphatic chains. Aromatic solvents, such as xylene (dimethyl benzene) and toluene (methyl benzene) are particularly useful in dissolving gums and resins and are therefore widely used in paints and lacquers. Some of the aliphatic and aromatic hydrocarbons used in formulations are shown in Figure 1.5.

Hydrocarbon Blends and Distillates

Many commercial hydrocarbons solvents are not supplied as pure substances composed of only one compound but are mixtures of related compounds. For most solvent applications a pure substance is no better at the job than a mixture; for example, it may be thought that an aliphatic hydrocarbon with a relatively high boiling point and reasonable evaporation rate, such as isoundecane, would be suitable for a particular task.

To opt for that particular substance as a pure chemical would be expensive, but a mixture of chemicals containing mainly isoundecane along with other members of the same chemical family such as the decane and dodecane isomers, and available at a fraction of the cost, would be just as effective.

Another example, and rather similar to the previous one, is hexane. A

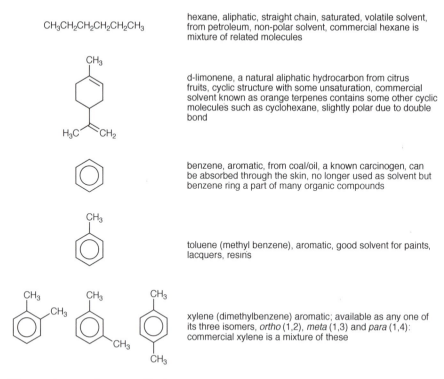

CH₃CH₂CH₂CH₂CH₂CH₃ hexane, aliphatic, straight chain, saturated, volatile solvent, from petroleum, non-polar solvent, commercial hexane is mixture of related molecules

d-limonene, a natural aliphatic hydrocarbon from citrus fruits, cyclic structure with some unsaturation, commercial solvent known as orange terpenes contains some other cyclic molecules such as cyclohexane, slightly polar due to double bond

benzene, aromatic, from coal/oil, a known carcinogen, can be absorbed through the skin, no longer used as solvent but benzene ring a part of many organic compounds

toluene (methyl benzene), aromatic, good solvent for paints, lacquers, resins

xylene (dimethylbenzene) aromatic; available as any one of its three isomers, *ortho* (1,2), *meta* (1,3) and *para* (1,4): commercial xylene is a mixture of these

Figure 1.5   *A wide range of hydrocarbons, both aliphatic and aromatic, are used in formulations. Most are from petroleum and a few from plant materials*

particular application may require hexane as a solvent but to purchase this as pure *n*-hexane costs many times as much as the distillate which is mainly *n*-hexane along with its isomers and small amounts of pentane and heptane.

Distillates commonly available are: odourless kerosene, white spirit, normal paraffins, iso-paraffins and aromatics. All are composed of a mixture of similar hydrocarbon molecules and, as such, have a range of boiling points rather than the single value expected of a pure compound. The aliphatic distillates such as white spirit are low cost solvents and are therefore widely used in paints and for cleaning/degreasing. Aromatic hydrocarbons are also available as distillates containing different hydrocarbons, for instance, commercial xylene contains the three xylene isomers.

Natural solvents are often mixtures composed of molecules that may be chemically quite different. For example, the citrus oil d-limonene is commercially available as orange terpenes which is made up of *para*-mentha-1,8-diene along with cyclohexane and related compounds.

Oxygenates

The various groups of oxygenates are: alcohols, ketones, ethers, acids, esters and combinations of these such as alcohol ethers. Second to hydrocarbons the oxygenates are common ones found in everyday formulations.

Because of the strong electronegativity of the oxygen atom the oxygenates are polar solvents. As such, many of them are attracted to water molecules which are also polar. The result is a strong interaction made up mainly of hydrogen bonds as a solution forms. For example the lower alcohols are infinitely soluble (miscible) in water. If 50 ml of ethanol is mixed with 50 ml of water a noticeable amount of heat is evolved and the final volume is 98 ml – strong evidence of powerful attraction between ethanol molecules and a water molecules.

As the carbon chain increases in length in the alcohol its water solubility becomes less because it has a greater proportion of non-polar hydrocarbon content. The result is that the hydrogen bonding role played by the hydroxyl groups diminishes. Of course, if the number of hydroxyl groups in the molecule is increased this will promote water solubility through increased hydrogen bonding.

Oxygenates with a variety of oxygen functional groups offer a combination of good solvency for grease and good water solubility. The glycol ethers are particularly important in this respect. Examples of commonly used oxygenates are shown in Figure 1.6.

Halogenated Hydrocarbons

Replacing hydrogen atoms in hydrocarbons with one or more of the halogens (fluorine, chlorine, bromine or iodine) results in halogenated hydrocarbons. Where chlorine is the halogen some very powerful and widely used solvents exist but unfortunately they tend to have high Ozone Depletion Potentials (ODP) and other undesirable environmental effects. Halogenated hydrocarbons, mainly chlorohydrocarbons, are denser than water and once in a watercourse, a river bed for instance, they can remain there for years as their evaporation is blocked by the water and they have very low water solubilities. Thus a watercourse burdened with these may be rendered lifeless for years, maybe hundreds of years.

There has been considerable recent development in the field of halogenated hydrocarbons resulting in the chlorofluorocarbons (CFCs) which cause ozone depletion by reaction in the stratosphere being replaced by hydrochlorofluorocarbons (HCFCs) which break down in

CH₃OH	methanol (methyl alcohol), traditionally obtained by destructive distillation of wood, poison, causes blindness, added to industrial ethanol to make methylated spirit
CH₃CH₂OH	ethanol (ethyl alcohol), important solvent, two sources; nature-made in fermentation and man-made from petroleum
(CH₃)₂CHOH	isopropanol (isopropylalcohol, IPA), another alcohol, common solvent, low cost, man-made from petroleum
CH₃OCH₃	dimethyl ether (methoxymethane), boils at minus 25 Celsius, used as aerosol propellant, extremely flammable
CH₃COCH₃	acetone (propanone), a ketone, common solvent in paints and lacquers, highly flammable
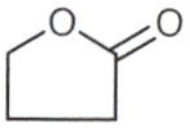	gamma butyro lactone, a ketone ether from lactic acid, good solvent for polymers and resins
CH₃COOC₄H₉	butyl acetate (butyl ethanoate), an ester, powerful solvent, man-made from petroleum, widely used in solvent based paints especially cellulose paints
CH₃(CH₂)₃O(CH₂)₂O(CH₂)₂OH	butyl Carbitol, (butyl digol, diethylene glycol monobutyl ether), used in aqueous cleaning formulations, other Carbitols in this commercial range
CH₃CH₂OCH₂CH₂OH	Cellosolve, 2-ethoxyethanol, resin and polymer solvent, other Cellosolves in this commercial range
COOCH₃ (CH₂)ₓ COOCH₃	where *x* = 3, 4 or 5; dimethyl esters of glutaric, adipic and succinic acids, an ester blend based upon natural chemicals, commercially available
CH₃CHCOOC₄H₉ | OH	butyl lactate, ester/alcohol from lactic acid

Figure 1.6 *Oxygenates are hydrocarbons containing oxygen in a variety of functional groups and represent a major category of solvents. Although most are synthesized from petroleum there is a significant number from natural processes*

the troposphere and are thus prevented from reaching the higher altitude of the stratosphere where the ozone reactions occur. Other replacements for CFCs are the hydrofluorocarbons (HFCs). Fluorinated compounds such as the hydrofluoroethers, (HFEs) are finding use as solvents but their high cost restricts them to highly specialized work such as medical processes. Figure 1.7 gives examples of common halogenated compounds.

CCl₃F	trichlorofluoromethane, CFC 11, ODP = 1.0, was used as aerosol propellent, now phased out
CHClF₂	chlorodifluoromethane, HCFC 22, ODP = 0.030, making plastic foam
CH₂FCF₃	1,1,1,2-tetrafluoroethane, HFC 134a, now used as aerosol propellent
CH₂Cl₂	dichloromethane, a common and powerful solvent used in paint stripper
ClCH=CCl₂	trichloroethylene (trichloroethene), excellent solvent for vapour degreasing of metals but hazardous to health
CH₃CH₂CH₂Br	l-bromopropane, relatively new on the scene and partly replaces the chlorine equivalents because it is not an ozone depletor

Figure 1.7 *Halogenated hydrocarbons represent some of the best solvents available but the main member of the group, the chlorine-containing members and the CFCs, are being replaced by other solvents due to their being implicated in damaging the ozone layer. ODP = Ozone Depletion Potential*

Nitrogen Compounds

Replacing a hydrogen atom in a hydrocarbon structure with a nitrogen atom gives another group of compounds in which there are some very useful solvents. Some of these are related to the base, ammonia, and exhibit basic properties such as an ability to react with acidic substances; for example, the reaction between oleic acid and monoethanolamine to form the soap monoethanolamine oleate.

Nitrogen compounds are used to scavenge traces of acid formed when trichloroethylene is used in degreasing baths. Many nitrogen compounds also behave as corrosion inhibitors for ferrous metals because the nitrogen atom can bond with iron atoms at the surface of the metal – see corrosion inhibitors. Some of the more widely used nitrogen compounds are shown in Figure 1.8.

Sulfur Compounds

Sulfur figures quite a lot in surfactants in the form of sulfates and sulfonates. In that form it is virtually odourless and non-toxic. In its other chemical groups such as thiols, sulfides and thiophenes it is poisonous and often has a repulsive odour. Thus, there are very few sulfur-containing solvents in formulation work. However, one notable member of this group is dimethyl sulfoxide $(CH_3)_2SO$, which is the sulfur analogue of acetone. It is a good solvent and does not have the stench of other sulfur compounds.

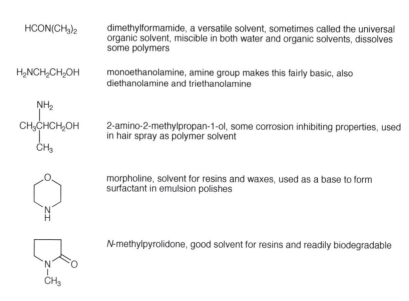

HCON(CH₃)₂ — dimethylformamide, a versatile solvent, sometimes called the universal organic solvent, miscible in both water and organic solvents, dissolves some polymers

H₂NCH₂CH₂OH — monoethanolamine, amine group makes this fairly basic, also diethanolamine and triethanolamine

CH₃CHCH₂OH (with NH₂ and CH₃ substituents) — 2-amino-2-methylpropan-1-ol, some corrosion inhibiting properties, used in hair spray as polymer solvent

morpholine, solvent for resins and waxes, used as a base to form surfactant in emulsion polishes

N-methylpyrolidone, good solvent for resins and readily biodegradable

Figure 1.8 *Nitrogen-containing organic solvents are frequently basic due the chemistry of the nitrogen atom. They are widely used in formulations, often because of this property*

Choice of Solvent

There is an infinite number of possibilities for solvent types because of the different arrangements of carbon atoms in a hydrocarbon along with a number of different chemical groups that can be attached to the hydrocarbon skeleton. From the above examples it is possible to form a general idea on what each group has to offer.

The large number of solvent types and properties means that solvent choice is going to require consideration of many factors. Of course solvency power is going to be high up the list but other factors may dominate. For example, acetone (propanone) may be perfect in terms of solvency power, and its miscibility with water, but its low flash point puts it in the highly flammable category.

In addition to the physical properties of solvents, account must also be taken of the potential for chemical reaction between the solvent and other chemicals present. Even a seemingly inert blend of oxygenates with chlorohydrocarbons can be an explosive mixture. There will also be many other considerations: evaporation rate; environmental effects from use, disposal and spillage; health and safety, for example, Control of Substances Hazardous to Health (CoSHH) Regulations; and cost.

WAXES

Waxes have a long history of use in formulations for waterproofing, polishing and cosmetics. These are organic substances from plants, animals or synthetic sources, the latter relying upon petroleum as the raw material. For the most part, the natural ones are long chain fatty acid esters or long chain fatty alcohols. Some notable exceptions are ones such as lanolin. In formulation work, waxes play a major role in the preparation of polishes of which there are three different types: pastes, solvent-based liquids and aqueous emulsions (wax/water and wax/solvent/water).

Animal Waxes

Beeswax

Ever since waxes were first used beeswax has held a position of prominence. It is composed of a mixture of esters such as myricyl palmitate and related compounds all of which have long chain fatty acids/alcohols along with small amounts of hydrocarbons. In common with most natural chemicals it is a complex mixture and is subject to wide compositional variation which depends upon its source.

It has low solubility in most solvents; in turpentine it is about 8% soluble but in white spirit only about 4% will dissolve. The wax is available in several grades depending upon purity and colour. Some of the compounds common to the various grades of beeswax are shown in Figure 1.9.

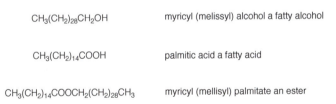

$CH_3(CH_2)_{28}CH_2OH$ myricyl (melissyl) alcohol a fatty alcohol

$CH_3(CH_2)_{14}COOH$ palmitic acid a fatty acid

$CH_3(CH_2)_{14}COOCH_2(CH_2)_{28}CH_3$ myricyl (mellisyl) palmitate an ester

Figure 1.9 *Most natural waxes, from animals or plants, are complex mixtures of long hydrocarbon chain fatty alcohols, acids and esters of these. The compounds shown here represent the main ones in beeswax but the relative amounts and the amounts of other similar compounds are dependent upon the source of the beeswax*

Shellac Wax

Produced by the lac insect, this wax is a complex mixture of the esters ceryl lignocerate, lacceryl laccerate along with lacceryl alcohol. It will dissolve in aqueous alkaline solutions because under these conditions some of the esters are saponified to form salts of aleuritic acid and shellolic acid which have soap like properties. Shellac wax is an important component of many polishes because of its ability to produce high gloss. French polish is based on shellac.

Lanolin

Sometimes known as wool-wax or wool-fat lanolin is from the sebaceous glands of living sheep and comprises a complex mixture of esters from over thirty different fatty acids and thirty different alcohols. Cholesterol, a steroid alcohol, is one of the main alcohols in the lanolin esters.

Lanolin has always held a firm place in the preparation of ointments because of its capacity to act as a vehicle for lipid soluble drugs to be administered to the skin. Normally lanolin contains about 30% water but anhydrous lanolin is also available. Derivatives of lanolin such as lanolin ethoxylates are manufactured for specific applications.

Plant Waxes

Carnauba

This comes from palm leaves and represents the most valued of the vegetable waxes. Available in different purity grades it is a hard, brittle, high melting point wax, and is highly rated for use in polishes where it is combined to advantage with beeswax, but it is expensive. It is composed of myrisyl cerotate, myricyl alcohol and related compounds.

Mineral Waxes

Ozokerite

Sometimes known as earth wax, ozokerite varies in colour in its crude form from green to black but is refined by bleaching to produce a yellow wax. The excellent solvent retaining properties of ozokerite make it a valuable material for blending with other waxes in the preparation of polishes.

Paraffin Waxes

Paraffin waxes and their microcrystalline relatives come from the refining of petroleum, as such they are not from renewable resources unlike the ones mentioned above. These waxes are mixtures of long chain hydrocarbons from twenty carbon atoms upwards. In the microcrystal-line waxes there is a large proportion of branched chain hydrocarbons.

Synthetic Waxes

Polyethylene Glycols (PEGs)

Unlike other waxes these are water soluble, producing clear colourless solutions. A range of these of different molecular weights, relating to the number of $-CH_2CH_2O-$ units per molecule, are available. Some-times known as Carbowaxes or PEGs each type is given a number indicating the approximate molecular weight. Thus, $HO(CH_2CH_2O)_{25}H$ is polyethylene glycol 1000.

This type of structure based on a chain of ethoxy (ethylene oxide) groups will be seen again in the surfactants section. An important prop-erty of PEGs is that of water solubility; the PEG waxes are completely soluble in water due to the large capacity for hydrogen bond formation resulting from the ethylene oxide groups. In effect the whole molecule can be hydrogen bonded to water molecules.

OILS

The term oil covers a large range of substances of different chemical types. In formulation work the most commonly encountered are: hydro-carbon oils (long chain aliphatics) such as the white oils used in skin care emulsions; the silicones, polydimethylsiloxanes and polymethylphenyl-siloxanes in preparation of polishes, barrier creams, release agents and anti-foams; vegetable oils (triglycerides) to make food-safe lubricants and cosmetics; essential oils such as terpenes used as bactericides, insect repellents, solvents and fragrances.

Silicones, which are entirely synthetic, offer a wide range of viscosities from water thin to a million times that of water and, unlike other oils, their viscosities vary little with temperature. They are also heat stable, odourless and colourless. The silicones are based on a silicon–oxygen (siloxane) chain with attached alkyl groups as shown in Figure 1.10.

Oils can be formulated as emulsions or dissolved in an appropriate solvent. In a few instances, particularly where essential oils are a key feature of the preparation, the oil may be solubilized by a relatively large quantity of surfactant.

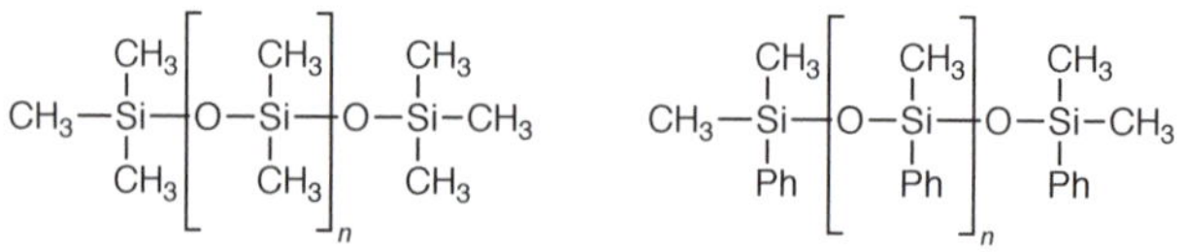

Silicone oils are synthetic poly(dimethylsiloxane) and poly(phenylmethylsiloxane) molecules with a viscosity range of 1 to 1,000,000 that changes little with temperature; they are thermally stable, odourless, colourless, inert and have a high flash point

Figure 1.10   *Silicones are synthetic fluids based on a siloxane chain rather than other oils where the chains are hydrocarbons*

ANTIOXIDANTS

The majority of formulations have to withstand lengthy periods of storage, sometimes in warm conditions along with intermittent exposure to the air. Under these conditions some ingredients are prone to oxidation and form undesirable oxidation products. This type of reaction can be catalysed by the presence of metal ions and these may dealt with by appropriate additives – see section on chelating agents. An example is the oxidation of unsaturated fatty molecules that turn into rancid smelling compounds, discolouration or break down of the product.

To offer protection against oxidation, antioxidants are added to preparations in small amounts to act as oxygen scavengers. Some examples are butylated hydroxytoluene (BHT), butylated hydroxyanisole (BHA) and propyl gallate and ascorbic acid. The latter is a natural antioxidant.

BIOCIDES

For the formulations considered in this book there are two groups of biocides: (1) preservatives to prevent spoilage and the transfer of harmful micro-organisms in, for example, shampoos; (2) disinfectants where the purpose of the product itself is to act as a biocide for, say, use as in bactericidal surface cleaner. Biocides are therefore intended to be toxic, designed to kill a target organism.

As such they are a risk to other organisms, including humans, and the environment. The need for regulation and control for this group of poisonous chemicals is obvious. There has been some confusion as to the actual definitions of the various groups of biocides and there are still differences from one country to another. However, in the EEC, the Biocidal Products Directive (Directive 98/8/EC) was introduced as an instrument for regulation of current and new biocides. An important feature of this legislation is that the amount of a biocide used in a formulation must comply with the stated concentration for that particular biocide.

It is common practice to use a combination of biocides to ensure effectiveness to kill a wide spectrum of micro-organisms. Where biocides are added as product preservatives the intention is to protect a product during its shelf life and once it is in the hands of the user. In personal care products the regulations require there to be sufficient biocidal preservative to destroy the micro-organisms: *Candida albicans*, *Pseudomonas aeruginosa* and *Staphyloccus aureus*. These are ones that, if allowed to reach high enough concentrations as may occur in a product with no biocide, could cause illness to the user.

Common biocidal preservatives for personal care products are: isothiazolines and the alkyl esters of 4-hydroxybenzoic acid, namely, methyl parabenz, ethyl parabenz, propyl parabenz, butyl parabenz.

The potential reactivity of a biocide once in a formulation is important as some biocides are rendered inactive by chemical reaction. For example, a cationic surfactant acting as a biocide may combine with an anionic surfactant. Even where a biocide is not deactivated by chemical reaction it may have its activity suppressed by other components of the preparation such as having a relatively large amount of non-ionic surfactant in which the bactericide can be trapped within the micelles – effectively isolated.

Preservatives kill the micro-organisms that cause the products to break down, *i.e.*, they prevent it from biodegrading. It seems a little contradictory that we are keen to find chemical ingredients that are biodegradable and once we have them we put in preservatives to stop them biodegrading. However, when such formulations enter the environment the preservatives are so inundated by micro-organisms that they soon lose their activity and biodegradation takes its normal course.

THICKENERS

Added to a liquid formulation these substances increase viscosity and change the flow characteristics. They have a particularly useful role to play as emulsion stabilizers. Natural substances, and their man-made derivatives, are well represented in this group and include: the polysaccharide gums, arabic, acacia, guar and xanthan; cellulose derivatives such as sodium carboxymethylcellulose (SCMC) and hydroxy propyl methyl cellulose; starches such as corn starch; proteins, for example gelatine and inorganic ones, clays, *etc*.

SCMC is highly effective and finds wide application. For example adding only 1% of this thickener to water increases the viscosity by a factor of two thousand. Synthetic thickeners such as the polyacrylate emulsions are more versatile than the natural ones and offer particular advantages such as remaining water thin for easy mixing and then having their thickening properties developed by pH adjustment and they are less prone to attack by micro-organisms than the natural thickeners. Some inorganic man-made thickeners such as fumed silicas are also available and find use in thickening of non-aqueous systems based on solvents.

In addition to actual thickeners some surfactant preparations may be thickened up simply by adding electrolytes such as common salt. This is discussed in Chapter 2. In a similar manner certain surfactants can be added as thickeners as is the case where sodium hypochlorite solution is made into thick bleach.

An interesting property of some thickeners is their ability to be thick until agitated, stirred or shaken, upon which they become thin. These are the thixotropic additives and strictly speaking should be referred to as rheology modifiers. Common examples of thixotropic preparations are non-drip emulsion paint and tomato ketchup; both these become thin due to the mechanical action of brushing on or of shaking the bottle.

CORROSION INHIBITORS

Whenever an aqueous preparation comes into contact with metals the potential for corrosion has to be taken into account. This is particularly so when the formulation is used in machinery especially where that use entails long term contact. The problem is added to by the presence of the range of metals and alloys that are typical of modern plant and machinery.

Preparations for putting into metal containers, such as aqueous aerosol products, must also be considered in terms of the potential for corrosion. Even non-aqueous formulations may contain traces of moisture that may lead to corrosion in the can. Metalworking formulations, lubricants and water in circulatory systems such as are used for heating and cooling rely heavily upon corrosion inhibitors.

In general a corrosion inhibitor is used in small concentrations in a preparation. The actual amount for a particular task must be given some serious thought because inhibitors become bonded to metal surfaces and are therefore gradually consumed so that the corrosion inhibiting capacity may be spent before the remaining chemicals have become exhausted.

When considering adding a corrosion inhibitor into a formulation it is important to take account of the fact that some inhibitors may well prevent corrosion of steel but enhance corrosion of other metals or alloys such as brass. Although inhibitors may prevent the onset of corrosion they have no effect on metals that already have surface corrosion – they are not, in general, corrosion remedial agents.

Common examples of corrosion inhibitors used in formulations are: amines, ethanolamines, ethanolamine borates, benzotriazole, *N*-acyl sarcosines, methylene phosphonic acid derivatives, sodium nitrite, sodium silicate. Quite a few of these are nitrogen compounds because of the ability of that atom to bond with atoms at a metal surface and effectively insulate the metal from the corrosive environment.

There is currently concern about nitrites as anti-corrosion agents because of the potential for them to react with amines to form *N*-nitrosamines which are carcinogenic. It seems ironic that, despite such concerns, sodium nitrite is added to some meat products such as canned pork and ham to prevent the growth of deadly bacteria and make the meat a healthy shade of pink.

FOAM MAKERS AND BREAKERS

As foam is a result of surface activity the theoretical aspects are considered in the chapter on surfactants (Chapter 2). In many applications the presence of foam is important whereas other applications may require no foam. Thus, substances that make foam as well as ones that break foam are on the list of chemicals used in formulation work.

The presence of foam in a shampoo or dish-wash liquid is probably as much a psychological feature as a technical asset. For many of us there is the perception that if the shampoo does not foam then it is no good. However, in some preparations foam is essential to the working of that preparation, for instance shaving cream would be near useless without the foam.

Different surfactants offer different foaming capacities and prudent choice of surfactant may produce the desired result without having to put foam booster into the formulation. Where it is necessary to use a foam booster the choice is more often than not coconut diethanolamide. (CDE):

$$R-\underset{\underset{O}{\|}}{C}-N\begin{cases} CH_2CH_2OH \\ CH_2CH_2OH \end{cases}$$

$R = C_{12}H_{25}$ to $C_{14}H_{29}$.

Surfactant preparations in many industrial processes encounter severe agitation or high pressure spraying with the consequence that foaming is a potential problem. Of course the use of low foam ingredients goes a long way to eliminating this. However, situations will still arise where anti-foaming agents are required.

ABRASIVES

Insoluble hard solids ground to provide a variety of particle sizes are added to preparations where an abrasive polishing action is required such as metal cleaners and toothpastes. Important aspects in the consideration of abrasives are: particle size, hardness, chemical reactivity, resource and environmental aspects. With regard to chemical reactivity a calcium carbonate abrasive would not be chosen where acidic components were to be incorporated in the preparation.

From environmental and resources viewpoints it is worth bearing in mind that most abrasives are of non-renewable mineral origin and the processing of these may not be the most environmentally friendly activities. Examples of common abrasives used in formulation work are: calcium phosphates, carborundum, china clays, precipitated silicas, calcium carbonate. For liquid preparations the powder should, ideally, be held in suspension, but because of the high density of these materials this presents difficulties and, thus, many abrasive formulations are made up as pastes.

COLOURS

In many everyday preparations colour is required and this is provided mainly by soluble dyes or, to a lesser extent, insoluble pigments. Thus for an aqueous formulation a water-soluble dye is chosen whereas for oils and solvents the oil-soluble dyes are used. For any colourant it is important to bear in mind that the colour may be highly dependent upon pH and chemical environment. Exposure of the formulation to ultraviolet light, particularly strong sunshine, may completely or partially destroy a colourant. In some instances UV absorbers have to be added to protect the colour.

Of the dyes there are the natural dyes, coal tar dyes and azo dyes and each has its own identifier known as its Colour Index (CI) number. Care must be taken before using a colourant in a preparation, especially one for personal care, to ensure that it complies with appropriate legislation. And it may be permitted in one country but banned in another.

FRAGRANCES

Fragrances are added to formulated products for two main reasons: to mask the odour of the chemicals in the product or to give the preparation an odour that has the right kind of associations for the consumer – certain products, particularly household ones, are expected to have certain smells. The fragrance used in a formulation can actually be the purpose for the user choosing that product in preference to the others that may even be better products.

Fragrant chemicals may also be the main active component in a formulation. For example an air freshener is bought because it releases fragrant chemicals into the atmosphere. Some of these spray products, it is claimed, actually eliminate odour-causing molecules. The suggestion that the offending substances just vanish is not quite accurate – even nasty smells are subject to the law of conservation of matter.

With aerosol air fresheners the offending odour molecules dissolve in the spray, precipitate with it and leave a chemical film on the surface upon which it settles. In most cases it is simply a matter of masking an unpleasant odour with a more acceptable one. As such the air is not purified by removal of smelly volatiles but is in fact made more polluted by the addition of extra chemicals – mainly mixtures of aldehydes, ketones, esters, terpenes and aerosol propellant.

Chapter 2

Surfactants in Action

It is difficult to imagine how contemporary industrial society could manage without surface active agents, surfactants. Used extensively in the home, industry and agriculture these substances are a development of soap which had its origins in days of antiquity. Being made from animal or plant fats and going back to a time when people lived in harmony with nature there is a temptation to think of soap as a natural substance. It is the result of a man-made chemical reaction as will be seen below and, as such, is not entirely natural but it does retain much of its nature-made chemical content.

In fact there are few natural surfactants to be found and those that are available are neither of sufficient abundance or do not have the desired properties to satisfy modern needs. For example there are foam forming and emulsifying saponins such as the extracts from the soap plant, California soaproot. Other natural surfactants are found in biological processes that require surface active properties; small amounts are made in the liver and form an active part of the bile salts, and the phospholipids of cell membranes have surfactant structured molecules.

From early times up to the present soap making expanded and developed to meet user demands, to exploit new raw materials and to respond to new manufacturing processes. Despite the huge growth in the modern equivalents to soap it is still the most widely used single surfactant. It is almost certain that the first soap was a chance observation that animal fat that had mingled with the hot ashes of a wood fire had some amazing properties when mixed with water, not least of which was an ability to create foam and help wash away dirt from greasy bodies and clothing.

Animal fats and vegetable oils are compounds of long chain fatty acids (higher carboxylic acids) and the trihydric alcohol glycerol (propane-1,2,3-triol) bonded by means of ester linkages to give triglycerides as shown in Figure 2.1.

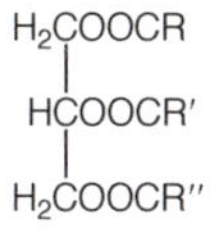

H_2COOCR
|
$HCOOCR'$
|
H_2COOCR''

where R, R' and R'' represent the long hydrocarbon chains of fatty acids

Important natural fatty acids in soaps and other surfactants are:-

A. Saturated hydrocarbon chains

C_{12} lauric (dodecyl) $CH_3(CH_2)_{10}COOH$

C_{14} myristic (tetradecyl) $CH_3(CH_2)_{12}COOH$

C_{16} cetyl (hexadecyl) $CH_3(CH_2)_{14}COOH$

C_{18} stearic (octadecyl) $CH_3(CH_2)_{16}COOH$

B. Unsaturated hydrocarbon chains

C_{18} one double bond
oleic, *cis*-9-octadecenoic

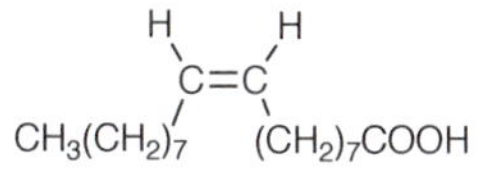

C_{18} two double bonds
linoleic, *cis,cis*-9,12-octadecadienoic

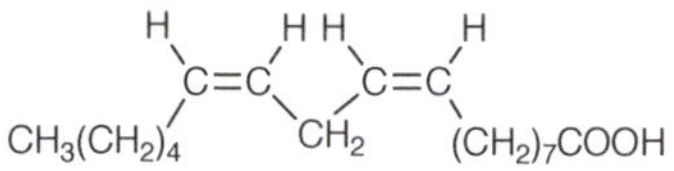

Figure 2.1   *Long chain fatty acids are found in combination with glycerol as the triglycerides which form the basis of animal fats and vegetable oils. In soap and surfactant chemistry the C_{12} to C_{18} fatty acids are the important ones. Some of the fatty acids are saturated, others are unsaturated with carbon–carbon double bonds*

Around fifty fatty acids, with saturated and unsaturated chains, occur in nature and all have an even number of carbon atoms. When the triglycerides are heated in the presence of alkali they undergo hydrolysis of the ester linkage followed by formation of the alkali metal salt of the fatty acids in a process known as saponification. Use of sodium hydroxide as the alkali produces a hard soap whereas potassium hydroxide gives a soft soap. Figure 2.2 shows the basic saponification reaction.

In addition to the formation of soap, the alkali metal fatty acid salt, there is also glycerol which may be left mixed in with the soap or separated and purified as a valuable by-product. Various soaps, hard or soft, animal or vegetable are produced industrially by the above reaction and variants of it. High temperature and high pressure reaction with water can bring about hydrolysis of the ester bonds and this process is now widely used in soap manufacture.

More recently, with soaps still going strong, has come the development

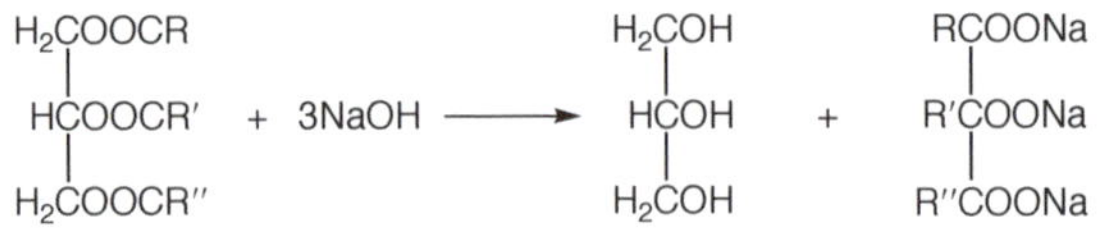

$$
\begin{array}{ccccc}
\text{H}_2\text{COOCR} & & \text{H}_2\text{COH} & & \text{RCOONa} \\
| & & | & & | \\
\text{HCOOCR}' & + \; 3\text{NaOH} \longrightarrow & \text{HCOH} & + & \text{R}'\text{COONa} \\
| & & | & & | \\
\text{H}_2\text{COOCR}'' & & \text{H}_2\text{COH} & & \text{R}''\text{COONa}
\end{array}
$$

$CH_3(CH_2)_{16}COONa$	sodium stearate
$CH_3(CH_2)_7CH:CH(CH_2)_7COOK$	potassium oleate
$CH_3(CH_2)_{16}CO_2NHCH_2CH_2OH$	ethanolamine stearate
$CH_3(CH_2)_{14}CO_2NH_4$	ammonium palmitate

Figure 2.2   *The saponification reaction hydrolyses the ester linkage and forms salts of the fatty acids. Sodium hydroxide is commonly used, in which case the sodium soaps are formed. Other soaps in which the fatty acid is neutralized with different bases give, for example, potassium soaps, ammonium soaps and ethanolamine soaps*

of other surfactants, many of which are very similar, in terms of their chemistry, to the soaps of millennia ago. It was during the First World War that the development of surfactants gained momentum to replace soaps which required animal fats and vegetable oils that were in short supply.

Of all the chemicals used in formulation work surfactants are by far the most important – except, of course, for water. Appearing in virtually every aqueous-based preparation, and some non-aqueous ones as well, surfactants carry out a variety of functions as will be seen later. There are many different chemical types of surfactants with molecular structures designed to carry out a particular job.

It is the capacity of surfactants to make water interact with other liquids (and solids) that explains the wide range of functions that form the basis for a large, and growing number, of formulated products. Presently the growth of surfactant-based preparations is very healthy due to environmental pressures forcing changes away from the use of volatile organic compounds (VOCs), as solvents, towards water-based systems. Two prime examples of this are: changes from solvent-based paints to aqueous-based ones and replacement of organic solvents in industrial degreasing processes with surfactant solutions.

To appreciate the versatility of surfactants, and to gain an insight into how they behave in different environments such as are to be found in formulations, requires a consideration of surface chemistry and surface tension.

SURFACE TENSION

It's like water off a duck's back. Fortunately for ducks surface tension effectively prevents water from soaking the animal's feathers because the oily coating on the barbs is strongly water repellent; the creature is thus saved the misery of being frequently drenched, cold and unable to fly due to excess baggage. Surface tension works just as impressively for the water spider that happily walks on the surface of the pond and does so without getting its feet wet.

Taking a close look at a water spider in action – or other water walkers – shows that the insect's feet actually cause the water surface beneath them to be pressed down as if the surface were in fact covered with a skin which becomes stretched under the weight of the insect. This is the clue to surface tension. The molecules at the surface are pulled tightly towards the bulk of the liquid with the effect that the surface becomes taut. Thus, surface tension may, for many purposes, be visualized as being like an elastic skin on the surface of water. Surface tension is not a property confined to water, all liquids possess it but there it is much weaker and less obvious than for water.

Although surface tension works in favour of the duck, the water spider and manufacturers of waterproof coatings it imposes difficulties when it comes to mixing certain liquids such as oil (or grease) and water. These mixing problems may, however, be overcome with the assistance of other substances that can change the surface tension. To see how this happens a consideration of some theoretical aspects is necessary.

Surface tension is explained in terms of the forces acting between the molecules in a liquid. Usually water is taken as an example because as this is where most surface tension effects of importance are found. Each molecule in the bulk of the liquid experiences attractive forces in all three dimensions from neighbouring molecules. A molecule at the surface of a liquid is deficient in the 'upwards' force but the strength of the other forces of attraction remains the same as shown in Figure 2.3. As a consequence there is a net downward pull that creates tension at the surface which manifests itself in the skin like properties.

Some substances when dissolved in water increase its surface tension whereas others have the opposite effect (Table 2.1). It is the latter that are most important in formulation work because reducing the surface tension of water enables it to mix with oil or grease as well as taking part in many other valuable functions.

Sodium chloride dissolved in water increases surface tension because of strong attractive forces between the water molecules and the sodium cations Na^+ and chloride anions Cl^-. This increased attraction between

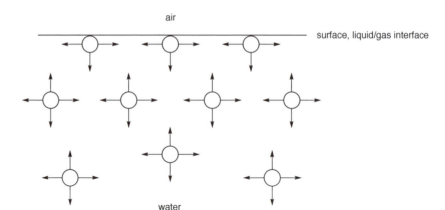

Figure 2.3   *Water molecules in the bulk are attracted to other water molecules, and with the same strength, in all three directions. Molecules at the surface of a liquid experience a net downwards pull that results in surface tension, which gives the impression of a surface skin*

Table 2.1   *Surface tension values, mN m^{-1} (dyne cm^{-1}), of pure compounds and aqueous solutions. Water has a much higher surface tension than the other pure liquids but its value can be changed by addition of different substances*

System	Surface tension (mN m^{-1})
Water	72
Water + 10% sodium chloride	76
Water + 10% methanol	59
Water + 0.1% surfactant	32
Octane	22
Ethyl acetate	24
Ethanol	23

Source: Handbook of Chemistry and Physics.

the particles of the liquid has the effect of making the surface tighter and that means a greater surface tension.

On the other hand, putting an alcohol like methanol into the water reduces the surface tension by weakening the attractions between water molecules due to the hydrocarbon part hindering the hydrogen bonds that are the major attractive forces between water molecules.

Most surfactants, at about the 0.1% level, will reduce the surface tension of water from 72 to about 32 mN m^{-1} (dyne cm^{-1}) but the highly specialized fluorinated surfactants can take it as low as 12 mN m^{-1}. It can be noticed from the above table just how much higher is the surface tension of water than that of other liquids. This explanation is because the molecules in most liquids are attracted to each other by weak Van der Waals forces whereas in water the attractive forces are hydrogen bonds which are much stronger. The stronger the intermolecular attractions within a liquid the greater is its surface tension.

The effect of the intermolecular attractive forces upon surface tension is also seen when water is heated. Heating water from 20 °C, where its surface tension is 72 mN m^{-1}, to boiling point results in its surface tension falling to 59 mN m^{-1} as the thermal energy of the molecules disrupts the hydrogen bonding.

Similar to the soap considered above, other surfactants have long hydrocarbon chains and the length of these, that is the number of carbon atoms in them, has an effect upon how good the surfactant is at reducing surface tension. For a surfactant molecule of a particular homologous series each additional CH$_2$ group makes the molecule about three times more effective at lowering surface tension.

To understand how surfactants work we need to consider in detail what their molecules are like. Surfactants are generally large molecules with two distinct parts: a hydrophilic (water loving) part and a lipophilic (oil loving) part. The hydrophilic part is sometimes referred to as the lipophobic part (oil hating) and the lipophilic part as hydrophobic part (water hating) part – these are not to be confused with lyophilic and lyophobic which mean solvent loving and solvent hating respectively. I prefer not to use the '-phobic' terms as they tend to imply repulsion and this is not the case. Molecules that have both lipophilic and hydrophilic properties are said to be amphiphilic; surfactant molecules are therefore amphiphiles.

Although there are many variations of surfactant molecules, and some of them quite complex ones, they can all be represented in the simple form shown in Figure 2.4 to understand how they function. The surfactant sodium lauryl sulfate consists of an anionic molecule with an accompanying cation. The latter is sometimes referred to as the molecule's counter ion.

ADSORPTION

When a surfactant molecule is in water its hydrophile is strongly attracted to the water molecules (polar molecules attract other polar molecules) whilst its lipophile (non-polar) experiences negligible attraction. Thus a conflict exists – there is an uncomfortable energy situation – and the surfactant molecule prefers a place on the surface of the liquid

(a)

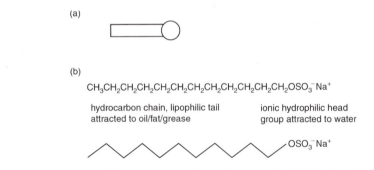

(b)

CH$_3$CH$_2$CH$_2$CH$_2$CH$_2$CH$_2$CH$_2$CH$_2$CH$_2$CH$_2$CH$_2$CH$_2$OSO$_3^-$ Na$^+$

hydrocarbon chain, lipophilic tail   ionic hydrophilic head
attracted to oil/fat/grease   group attracted to water

OSO$_3^-$ Na$^+$

Figure 2.4   (a) *General representation of a surfactant molecule showing the long lipophilic (hydrophobic) tail and relatively small hydrophilic (lipophobic) head.* (b) *Example of a common surfactant molecule, sodium lauryl sulfate*

where its lipophile orients itself away from the water. This concentrating effect of molecules at a surface is known as adsorption and it is fundamental to an understanding of surfactant behaviour, the basis of much formulation work. As a consequence of this adsorption it takes only a minute amount of surfactant to have a significant effect in terms of surface tension – see Figure 2.5.

No matter how small the concentration of surfactant there will always be some molecules present in the bulk of the water because of the dynamic aspects in which molecules are in constant motion resulting in an equilibrium between those in the bulk and the adsorbed ones. There is never a static situation because the liquid always has enough thermal energy to keep the molecules in a state of constant agitation as is demonstrated by Brownian motion experiments.

The surface of the water at which surfactant molecules adsorb is

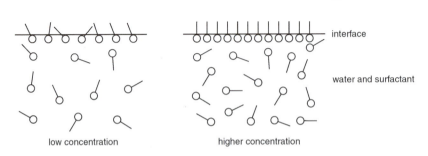

Figure 2.5   *Surfactant molecules adsorb on the surface of water (water/air interface) with their lipophiles oriented away from the water. At low concentrations orientation is random but becomes more ordered as more molecules are available at the higher concentrations*

simply a water-to-air interface (liquid/gas interface). Adsorption also occurs at other interfaces: solid/gas, solid/liquid, liquid A/liquid B. An example of the latter is the interface between oil and water and surfactant molecules can adsorb here with their hydrophiles in the water and their lipophiles in the oil as would be expected. However, adsorption of surfactant molecules at the water/solid interface depends on the type of surfactant molecule being adsorbed and the nature of the surface of the solid. Some surfaces are lipophilic, in which case the lipophile of the surfactant molecule is attracted. On the surface of some solids there is a positive or negative charge; this also affects the way in which a surfactant molecule is adsorbed (Figure 2.6).

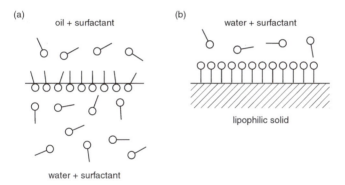

Figure 2.6   *Adsorption of surfactant molecules* (a) *at a water/oil interface,* (b) *on a lipophilic surface*

MICELLES

If we take a volume of water and progressively add more and more surfactant the surface tension of the water will gradually reduce until the point is reached at which the surface is packed full of surfactant molecules (Figure 2.7).

At this point the surface tension is at a minimum for the particular surfactant and beyond that point there is no further reduction. Once covered with an adsorbed monomolecular layer of surfactant molecules, the surface has, in effect, changed from a water surface (high surface tension) to a hydrocarbon surface (low surface tension). The random style of orientation of the molecules at low concentration is replaced at higher concentrations by a vertical packing until the surface is saturated.

But what happens as the concentration of surfactant molecules increases and the surface can accommodate no more? Clearly, these extra molecules must be accommodated in the bulk of the water but their low

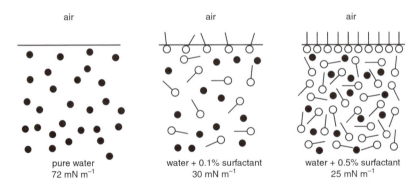

Figure 2.7   *As more surfactant molecules populate the surface then that surface changes from one of pure water with a high surface tension to one of part water part lipophile with lower surface tension to a surface which is completely lipophile and has minimum surface tension for the particular surfactant. The overall effect is one of changing a water surface to a hydrocarbon surface*

solubility allows for very few to exist as individuals. The majority are forced to agglomerate to form larger particles known as micelles. Surfactant micelles are best understood by considering the behaviour of an anionic surfactant (see later).

Evidence for the formation of micelles comes from studies based on surface tension, turbidity, osmotic pressure and electrical conductivity of aqueous surfactant solutions. In each of these properties, when a certain surfactant concentration is reached, a discontinuity is found.

This suggests that the molecules in solution no longer exist as discrete particles but start to associate and form aggregates of colloidal particles. The concentration at which aggregation occurs is known as the critical micelle concentration, CMC. As the lipophilic section of the surfactant chain increases in length its water solubility gets less and this corresponds to a lowering of the CMC. Thus the lower the solubility of surfactant the lower the CMC and *vice versa*.

Micelles have been identified as having essentially three distinct structures each of which depends upon the proportion of surfactant molecules to water molecules. The unassociated surfactant molecules are often referred to as monomers to distinguish them from the micelle forms. The simplest micelles are spherical aggregates of surfactant molecules with the lipophilic tails to the inside of the sphere and the hydrophilic heads to the outside and attracted by the water molecules as shown in Figure 2.8.

Surfactant molecules are arranged in these spherical micelles such that the lipophiles radially converge and the tail ends meet to give an approximate diameter of twice the lipophile lengths. Orientation of the

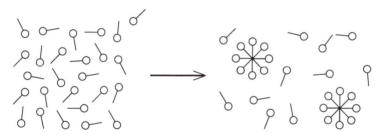

Figure 2.8 *Unassociated surfactant molecules form colloidal aggregates once a certain concentration, known as the critical micelle concentration, is reached*

molecules in a micelle is accounted for in terms of molecular attractions. Thus, lipophilic tails are attracted to each other by Van der Waals forces whilst the anionic head groups repel as a result of the ionic charges as shown in Figure 2.9.

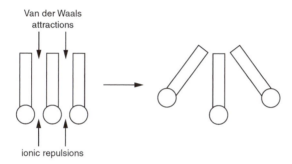

Van der Waals
attractions

ionic repulsions

Figure 2.9 *Attractive forces between the lipophilic tails along with the repulsions between the hydrophilic head groups have the combined effect of forcing surfactant molecules into a curve*

There is no hole in the centre of the sphere as the core is completely filled with lipophile. The number of molecules involved in a spherical micelle, known as the aggregate number, depends upon the particular surfactant and the temperature of the aqueous system. Aggregate numbers are reported in the specialized literature for a whole range of surfactant molecules. A typical anionic surfactant, on going from about 20 to 50 °C decreases in micelle size from about 70 to 40 molecules. Aggregate numbers in the larger spherical micelles can run into the thousands.

The micelles are not in a static situation and an equilibrium exists between surfactant molecules in the spheres and the unassociated ones in the bulk of the water. Taking account of the surface adsorbed molecules we can visualize a constant to-and-fro as molecules transfer between bulk, surface and micelle.

As the proportion of surfactant to water increases well above the

CMC the spherical micelles accommodate the excess by becoming elongated to form cylindrical rods (sphere-to-rod transitions) as shown in Figure 2.10. A solution composed of spherical micelles is slightly more viscous than one composed of unassociated surfactant because there

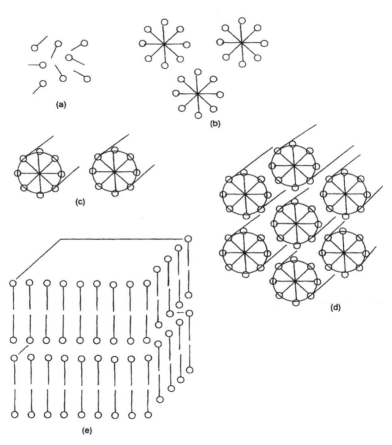

Figure 2.10   *Micelle formation as the concentration of surfactant molecules in water increases. (a) At very low concentrations individual molecules are present. Viscosity is about same as water. (b) At higher concentrations the molecules become associated into spherical micelles. Viscosity is about 100 times that of unassociated solution. (c) As the concentration of surfactant increases, with a consequential decrease in the number of water molecules the spherical micelles become cylinders. Viscosity is about 1000 times that of the unassociated solution. (d) With high concentrations of surfactants and decreasing water concentrations the cylinders form hexagonal arrangements. Viscosity is about 10,000 times that of unassociated solution. (e) In the neat surfactants where the water concentration is zero the surfactant molecules are arranged as layers. The viscosity here is low due to repulsions between facing anionic head groups*

is some point contact between the spheres, but as this sphere-to-rod transition takes place the cylinders so formed have a line of contact resulting in a big increase in viscosity.

Further increases in the amount of surfactant lead to a situation in which water molecules are becoming less available to occupy the spaces between the cylinders; the result is that the cylinders are forced into a close packing hexagonal arrangement. In this arrangement there are still sufficient water molecules to be attracted to the polar hydrophilic heads of the surfactant molecules.

Going to the final stage, where there is almost neat surfactant, the structure changes yet again. At these concentrations, when there is but little water available to participate in the structure, the micelle is lamellar with the individual surfactant molecules forming a palisade in which the lipophilic tails are in a layer structure. The opposing hydrophilic heads repel each other thus enabling the layers to slide freely which is reflected in a reduction of viscosity. As the degree of order increases once the hexagonal packing begins the micellar structures become liquid crystals.

The above situation is temperature dependent; below a certain temperature, known as the Krafft temperature, surfactant exists as hydrated crystals and unassociated molecules and no micelles.

The presence of electrolytes in the aqueous phase can have a profound effect upon micelle structure. For example a dilute aqueous solution of our typical anionic surfactant will have little viscosity because the surfactant is present as spherical micelles and unassociated molecules but the addition of sodium chloride dramatically increase the viscosity as the sodium and chloride ions force the hydrophilic head groups to pack closer together in their micelles. This results in a lowering of the CMC causing more aggregation. Depending on the amount of surfactant present this then leads to more sphere-to-rod transition and more hexagonal packing with a simultaneous viscosity increase. Further addition of salt eventually leads to thinning as the change towards the lamellar structure takes place due to the hydrophilic head groups being forced to pack together.

The opposite to this electrolyte thickening effect is found where a second surfactant with a very short lipophile is present and becomes wedged into the micelle, preventing it undergoing the sphere-to-rod transition and thereby stabilizing the spherical arrangement. Surfactants such as this are known as hydrotropes (see later) and keep the viscosity down even with dissolved salts and surfactants at quite high levels.

Much of what has been described above has been based upon systems of identical surfactant molecules. In reality this is hardly ever the case because commercial surfactants are a mixture of molecules with different lengths of lipophilic chains. Furthermore there may be residues from reactants and side reactions in the manufacturing process. As such, mixed micelles are the norm.

In formulation work mixed micelles offer much better performance in, for example, surface tension reduction and for producing stable emulsions. By means of two carefully chosen surfactants in the correct ratio a strong synergistic effect can be obtained. As such many formulations have two different surfactants arrived at by a process of optimization. Synergy of this type plays an important role in getting the best possible formulation from the minimum concentrations of chemicals.

TYPES OF SURFACTANTS

Surfactants are split into groups depending upon the nature of their hydrophilic head groups: anionic, cationic, non-ionic, amphoteric (Figure 2.11). As with all systems of classification there are grey areas and some may transcend the boundaries. For example, amine ethoxylates are generally non-ionic if the lipophile is a long chain and cationic if it is a short chain. Also, amphoteric surfactants can change to behave as cationics depending upon pH.

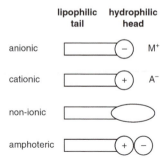

Figure 2.11 *Main types of surfactant with simplified representation of the lipophile and hydrophile*

Anionic Surfactants

Anionic surfactants are so called because the lipophile has a head group that carries a negative charge (Figure 2.12). Soaps are anionic surfactants based on a long fatty chain, as the lipophile, attached to a carboxyl group bearing a negative charge. This negative charge is countered by a positive ion, called the counter ion, which is usually a sodium ion but sometimes potassium or ammonium.

Although very similar to sodium lauryl sulfate, sodium stearate, has

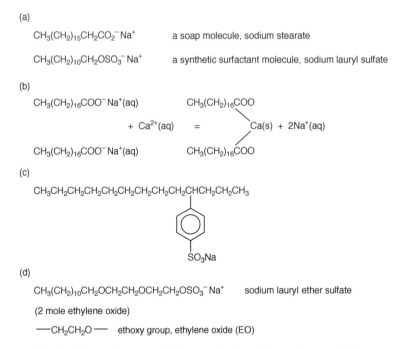

(a)

$CH_3(CH_2)_{15}CH_2CO_2^- Na^+$ a soap molecule, sodium stearate

$CH_3(CH_2)_{10}CH_2OSO_3^- Na^+$ a synthetic surfactant molecule, sodium lauryl sulfate

(b)

$CH_3(CH_2)_{16}COO^- Na^+(aq)$ $CH_3(CH_2)_{16}COO$

        $+ Ca^{2+}(aq)$ $=$ $Ca(s) + 2Na^+(aq)$

$CH_3(CH_2)_{16}COO^- Na^+(aq)$ $CH_3(CH_2)_{16}COO$

(c)

$CH_3CH_2CH_2CH_2CH_2CH_2CH_2CH_2CH_2CHCH_2CH_2CH_3$

$\overset{|}{SO_3Na}$

(d)

$CH_3(CH_2)_{10}CH_2OCH_2CH_2OCH_2CH_2OSO_3^- Na^+$ sodium lauryl ether sulfate

(2 mole ethylene oxide)

$-CH_2CH_2O-$ ethoxy group, ethylene oxide (EO)

Figure 2.12 *Anionic surfactants. (a) Some synthetic surfactants have a similar structure to soaps in that they are composed of a long fatty acid chain, a negatively charged head group and a positive ion such as sodium. (b) Sodium stearate, common soap, in water reacts with the calcium ions from the dissolved lime to produce calcium stearate (octadecanoate) that is insoluble, comes out of solution and is seen as scum. (c) Sodium dodecylbenzene sulfonate consists of a benzene ring with a long straight hydrocarbon chain as the lipophilic tail. (d) Ethoxy groups (ethylene oxide units) increase the hydrophilic strength of anionics*

one big disadvantage in that it reacts with lime, dissolved calcium salts, in the water to form calcium stearate which is insoluble and comes out of solution as scum. Thus, in hard waters much of the soap is lost as scum and so wasted. Of course, the water that is left behind after the scum has formed is then soft but the addition of soap as a means of softening water is an expensive practice. Besides, there is the problem of the scum floating around and sticking to whatever is being washed and, if clothes are being washed, leaving them with a dull or greyish appearance.

Sodium lauryl sulfate, in keeping with other synthetic surfactants, does not react to produce insoluble salts with calcium and so does not suffer loss of activity in hard water. In fact some synthetic surfactants are used in the hardest water of all, sea water, for cleaning up oil spills. It was the effectiveness of surfactants in hard water that provided much of the impetus in the early stages of development.

Soap also has the disadvantage that it works only in alkaline conditions. In general this requires the pH to be greater than 8. Putting soap into acidic aqueous conditions results in formation of the insoluble fatty acid and a situation similar to scum formation. The synthetic anionics can be used over wide pH ranges but some, particularly the fatty alcohol sulfates, can break down at low pH values, generally below pH 4, due to hydrolysis of the lipophile to hydrophile bond.

Synthetic anionic surfactants began with the sulfation of plant oils in the mid-nineteenth century. A major success was when the oil used was castor oil. The product found use in the dyeing of material with Turkey Red dye and from then on the sulfated castor oil was known as Turkey Red Oil. Other anionic surfactants, sulfates and sulfonates, were developed and made commercially available.

Later on, the sodium salts of alkyl benzene sulfonic acids were introduced into the commercial market. The first ones made had a branched hydrocarbon chain attached to the benzene ring. Although excellent as surfactants they posed environmental problems in that they were very slow to biodegrade and survived throughout the sewage treatment process. The outcome was that they continued their surfactancy in the watercourses into which the sewage effluent flowed. Rivers covered with foam along with masses of foam blowing around sewage plants were a common feature in the 1950s. And where water was taken from the rivers for the water supply there were cases of the water foaming whenever the tap was turned on – not a pleasant thought when one considers that the foam was a short time before in the sewage.

The problem of biodegradation had to be addressed and this was the driving force to develop anionic surfactants that would completely break down in the sewage treatment works. Success came in the form of straight hydrocarbon chain equivalents, linear alkylbenzene sulfonates (LABS), and the foaming rivers problem soon became history as these new anionics were more digestible to the bacteria that bring about biodegradation.

Nowadays many anionics not only contain sulfate or sulfonate groups to provide the hydrophilic part but also have ethylene oxide groups which enhance the attraction to water. Thus we have ethoxylates such as ether sulfates. Sodium lauryl ether sulfate with three ethylene oxide groups is a common component of modern shampoos and other personal care products, in which it represents the primary surfactant. It has excellent degreasing properties and is a good foam producer but to maximize its effectiveness it is often combined with other surfactants to take advantage of the synergistic effect.

The option of being able to add ethoxy groups, in any number, to a surfactant molecule offers a great deal of flexibility and enables the surfactant manufacturer to make a whole range of anionics with different levels of hydrophilic capacity. In the section on non-ionic surfactants it will be seen that the ethoxy group plays an even greater role and gives rise to a vast number of surfactant options and the benefit of making a surfactant which is tailor made for a particular purpose.

Anionics are widely used for their detergent action and, as such, are often the primary surfactant in preparations where detergency – cleaning power – is the function of the product. The anionics, because the hydrophilic head is relatively bulky, tend not to form micelles particularly easily and will remain as unassociated molecules in solution until concentrations reach percentage amounts before micelle formation begins. Hence their relatively high CMC values when compared with, say, the non-ionics discussed below.

This is advantageous in detergency and foaming because monomer surfactant molecules diffuse rapidly and, as a result, are always available when needed. On the other hand, aggregated molecules in the form of micelles are much slower. It is because of factors like these that anionics excel in detergency and foaming. However, although a high CMC is good for detergency it is also associated with skin and eye irritancy.

Another advantage of anionics in detergency is the anionic head enables the molecule to be electrostatically attracted to solid particles of soil thus facilitating their removal. Anionic surfactants are not to be used in conjunction with cationic surfactants (see below) as they contain opposite charges that would result in them being attracted and chemically combining with subsequent loss of activity of both.

Many domestic applications, washing-up liquids, shampoos, shower gels and the like rely upon anionics such as sodium lauryl ether sulfate as the main surfactant for cleaning action. In industry, anionics find a range of applications: hydrotropes (short chain lipophiles), wetting agents, foaming agents, detergents, emulsifiers for oils/waxes/solvents and emulsion polymerization, pigment dispersion, metal working, textile dyeing, pesticide emulsifiers and many more.

Apart from the chemistry of the hydrophilic head the length of the carbon chain representing the lipophile is also important from an application point of view. Generally, surfactants have lipophile chain lengths from between 8 and 20 carbon atoms. The short chain surfactants are fairly soluble in water and the long chain ones being less soluble as would be expected from the hydrophobic nature of the lipophile.

The smaller molecules make good hydrotropes whereas the larger molecules are better at detergency. A rough indication is that the C_{12} to C_{20} members give optimum detergency. At C_{12}, and less, optimum wetting and foaming characteristics are found. For each additional $-CH_2CH_2-$ solubility decreases and the CMC drops by a factor of ten.

Anionic surfactants are not just confined to the sulfates, sulfonates and soaps discussed above as will be seen from the following lists. The applications shown are merely by way of example and are not intended to imply that a surfactant has only that application. Most surfactants have a wide spectrum of functions and often are formulated into very different products.

Anionic surfactants	*Applications*
Alkylaryl sulfonates	
Sodium dodecylbenzene sulfonate	Hand dish-wash liquids
Calcium alkylbenzene sulfonate	Pesticide emulsifier
Triethanolamine alkylbenzene sulfonate	Car shampoos
Monoisopropylamine dodecylbenzene sulfonate	Solvent emulsifier
Sodium diisopropylnaphthalene sulfonate	Metal cleaners
Calcium alkyl aryl sulfonate	Aeration control in concrete
Alcohol sulfates	
Sodium lauryl sulfate	Shampoo, toothpaste
Lithium lauryl sulfate	Carpet shampoo
Triethanolamine lauryl sulfate	Fire fighting foam
Sodium ethylhexyl sulfate	Electroplating
Monoethanolamine lauryl sulfate	Bubble bath
Sodium tetradecyl sulfate	Pharmaceuticals
Ether sulfates	
Sodium lauryl ether (2EO) sulfate	Shampoo, bubble bath
Ammonium alkyl ether sulfate	Plaster board production
Ammonium lauryl ether (3EO) sulfate	Shampoo concentrate
Sodium alky lauryl ether sulfate	Polymerization
Phosphate esters	
Diethanolamine cetyl phosphate	Cosmetics emulsifier
Sodium lauryl ether phosphate	Electroplating
Acid alkyl phosphoric ester	Cleaning motors, machinery
Triethanolamine complex phosphate ester	De-inking

Sulfosuccinates

Sodium sulfosuccinate	Upholstery, carpet cleaner
Sodium dioctyl sulfosuccinate	Wax, oil emulsifier

Sarcosinates

Sodium lauroyl sarcosinate	Corrosion inhibitor
Ammonium lauroyl sarcosinate	Rug shampoo

Miscellaneous

Paraffin, olefin and petroleum sulfates/sulfonates
Taurates and isethionates
Carboxylates

Cationic Surfactants

Cationic surfactants another type of ionic surfactant. Here the lipophile is attached to a hydrophilic head that carries a positive charge, hence the name cation, with an anion as its counter ion. The structure can be thought of as being derived from a quaternary ammonium ion in which four groups, in this case four alkyls, are attached to a nitrogen atom which carries the positive charge. The detergency and emulsifying capacity of cationics is poor and they tend to be relatively expensive but, because of their particular structure, they have an important role to play.

A typical cationic surfactant is shown in Figure 2.13 from which it can be seen that out of the four alkyl groups attached to the nitrogen atom one is a long chain hydrocarbon which is clearly going to be the lipophile and the part that is attracted to oily substances.

What is most important in terms of application of these cationics is that the positive charge on the nitrogen atom is attracted to materials that carry a negative charge. Since the surfaces of most materials carry negative charges this gives rise to some interesting results when the cationic molecule, in aqueous solution, encounters such a surface.

When a cationic molecule with its positive charge is attracted to a material which has chemical groups carrying negative charges there is a strong ionic attraction which holds the surfactant molecule in place at specific chemical sites. The cationic surfactant undergoes chemical adsorption (chemisorption) as opposed to the type of adsorption previously considered which is physically adsorption (physisorption). A cationic surfactant is often referred to as being substantive to the surface. This chemisorption results in the properties of the surface being changed.

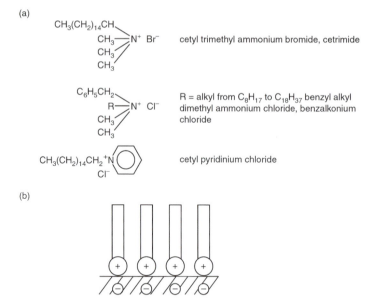

Figure 2.13   (a) *Cationic surfactant molecules normally consist of four alkyl groups (hence the name quaternary), representing the lipophilic part, attached to a positively charged nitrogen atom as the hydrophile. (b) Adsorption of a cationic surfactant molecule at negatively charged sites. Many materials, such as hair, skin, fibres and car paintwork carry negative charges when wet and attract cationic surfactants*

A familiar example of this is in hair conditioner where a cationic surfactant becomes chemisorbed to negative groups in the protein (keratin) that forms the hair shaft. The result: the protein becomes coated in cationic surfactant in which the lipophilic tails orient away from the surface and, in effect, change the surface from a protein one to a hydrocarbon one. Thus, a hitherto ionic and hydrophilic surface is rendered hydrophobic.

As such, the hair, after just being washed and dried, does not suffer the static electricity phenomena that causes fly-away hair when combed. Also, hair treated with cationic surfactant is softer and silky and has improved gloss. Cationic surfactants are therefore widely used in hair care products.

A similar effect is found when cationic molecules adsorb onto the skin, a protein, which then acquires a pleasing softness. Cationic surfactants therefore find use in skin preparations. Fabric conditioners work in the same way and make the fibres of the fabric softer. In addition to imparting softness the conditioned surface is easy to iron as it then has a certain amount of lubricity.

Some traffic film removers and car shampoos contain cationic surfactants which chemisorb to the vehicles paintwork because the polymeric structure of that paint or top lacquer carries negative charges. Thus, after washing with a preparation containing an appropriate cationic surfactant the paintwork is left with a hydrophobic layer which gives gloss and makes the water bead off. Although the cationic molecules are chemisorbed to the paintwork the effect is not as permanent as one would wish because of the quite harsh environment that vehicles encounter.

Because cationic surfactant molecules react with the negative sites on proteins to form stable compounds they are capable of disrupting biological systems and it is this that enables cationics to be used as bactericides. Their destructive action upon bacteria derives from the way they bond to the proteins of the cell membrane and prevent it functioning, causing the bacteria to perish.

Different cationics have different bactericidal properties depending on the groups in the quaternary structure. Bactericides based on these cationics, sometimes called quats (from quaternaries), often comprise a few different types to give a good spectrum and kill a range of bacteria. Single cationics can be quite selective in the bacteria that they attack. In general the damaging biological action is confined to killing bacteria but it can intrude into other biological systems and be a problem. If cationic surfactants get into the eye they adsorb onto the lens, made of protein, and can cause fogging. And ingested cationic surfactants can disrupt intestinal bacteria.

The properties of cationic surfactants take them into many other applications: road surfacing, mineral ore floatation, pigment dispersion, anti-statics, anti-caking agents. Wherever cationic surfactants are used it must be borne in mind that they can combine with most substances carrying a negative charge, including anionic surfactants.

When a cationic surfactant molecule is attracted to an anionic surfactant molecule the result is that both hydrophiles bond together, lose their activity and form a large molecule that is all lipophile – a fatty insoluble substance results. Caution therefore needs to be exercised whenever the addition of a cationic surfactant is considered as an additive to a preparation. However, in the field of hair care products there are some substances with cationic properties that can be combined with anionics.

Below is listed a few examples of cationic surfactants and uses. The list is not intended to imply that these cationics are confined to one particular function.

Quaternary ammonium cationics	*Applications*
Distearyl dimethyl ammonium chloride	Fabric softener
Lauryl trimethyl ammonium chloride	Shampoos
Cetyl trimethyl ammonium chloride	Algaecide, bactericide
Alkyl dimethyl benzyl ammonium chloride	Food industry bactericide
Alkyl trimethyl ammonium methosulfate	Hair conditioning
Coco trimethyl ammonium chloride	Pharmaceuticals
Cetyl pyridinium chloride	Mouthwash
Didecyl dimethyl ammonium chloride	Fungicide

Non-ionic Surfactants

Although the anionic surfactants discussed earlier play a major role in surfactant technology they are now almost equally matched by the non-ionic surfactants in which, as the name implies, there is no ionic character. Non-ionic surfactants arrived on the scene much later than the anionics but have grown to be major players as detergents and emulsifiers.

Because of their excellent emulsifying properties and the wide variety of different molecular structures, non-ionics form the core of emulsion chemistry acting as emulsifiers in for both oil-in-water emulsions and water-in-oil emulsions (see later). Nowhere is this best exemplified than in the manufacture of cosmetic and toiletry creams and lotions. The study of emulsion technology focuses largely on the non-ionic surfactants and a scientific approach for deciding upon the best ones for a particular emulsifying task has been developed, the hydrophile/lipophile balance (HLB) system. This is discussed in a later section.

Non-ionic surfactants can be manufactured from a wide range of raw materials amongst which many are plant based and therefore have the advantage of using renewable resources. In the examples that follow the lipophile is the hydrocarbon chain, just as with the ionic surfactants. Most of the non-ionics have as their hydrophile a series of ethylene oxide groups in the form of an ethoxylate chain. The more ethylene oxide groups present, the more hydrophilic the molecule, the greater its solubility in water. An ethoxylate with only two ethylene oxide groups will have virtually no water solubility but instead will be oil soluble whereas one with twenty such groups will have good water solubility but poor oil solubility.

This basic structure of a lipophile attached to an ethoxylate chain represents the most important class of non-ionics, the fatty alcohol ethoxylates and alkyl phenol ethoxylates. Usually in the naming of these substances the number of moles of ethylene oxide per mole of lipophile is indicated by the number followed by EO.

The hydrophilic ethoxylate chain in these non-ionic surfactant molecules attracts water molecules over the whole of its length by means of hydrogen bonding as shown in Figure 2.14. This is in contrast to the ionic surfactants where polarity is much more concentrated and attraction to water molecules is confined to the head group. Thus the ethoxylate hydrophile, in an aqueous environment, becomes sheathed in hydrogen bonded water molecules at each ethylene oxide group resulting in a bulky cylinder of hydration.

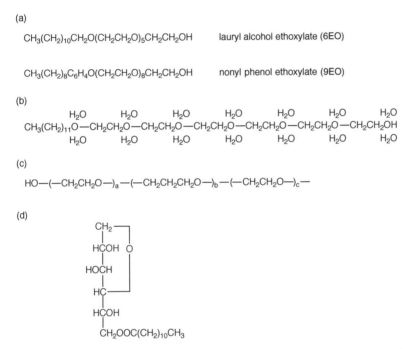

(a)

$CH_3(CH_2)_{10}CH_2O(CH_2CH_2O)_5CH_2CH_2OH$ lauryl alcohol ethoxylate (6EO)

$CH_3(CH_2)_8C_6H_4O(CH_2CH_2O)_8CH_2CH_2OH$ nonyl phenol ethoxylate (9EO)

(b)

(c)

$HO-(-CH_2CH_2O-)_a-(-CH_2CH_2CH_2O-)_b-(-CH_2CH_2O-)_c-$

(d)

Figure 2.14   *Non-ionic ethoxylates:* (a) *lauryl alcohol ethoxylate made by ethoxylating plant derived fatty alcohol and nonyl phenol ethoxylate, which requires petrochemical based raw material for ethoxylation;* (b) *cylinder of hydration around an ethoxylate;* (c) *entirely synthetic non-ionics made by copolymerizing ethylene oxide (EO) and propylene oxide (PO) to give EO/PO copolymers;* (d) *sorbitan esters are derived from natural renewable raw materials*

An interesting property of non-ionic ethoxylates dissolving by means of hydrogen bonding is that they come out of solution on heating as the thermal energy becomes sufficient to break these hydrogen bonds and reduce solubility. This is of course contrary to what is normally expected in that solubility increases with increasing temperature. Thus on heating a solution of non-ionic ethoxylate a temperature is reached at which point the solution becomes cloudy (known as the cloud point). As detergents are often required to be used in hot water this may seem a disadvantage. However, when these non-ionic ethoxylates are at elevated temperatures, but just below their cloud points, they are particularly effective.

The presence of dissolved salts also lowers the cloud point because ions from the salt compete for water molecules that hydrogen bond with the ethoxy groups in the hydrophile. Cloud points vary from one non-ionic ethoxylate to another depending on the relative lengths of the lipophiles and hydrophiles. For example, a long ethoxylate chain gives the molecule a high water solubility, as would be expected, and this is seen in the high cloud point (Table 2.2).

Table 2.2 *The greater the number of moles of ethylene oxide in the molecule the greater is the water solubility and the higher the cloud point for the same lipophile; the presence of an electrolyte such as sodium chloride reduces both solubility and cloud point because the ions compete for the water molecules in the hydration layer*

1% aqueous surfactant solutions	*(cloud point) (°C)*
Alcohol ethoxylate 6EO	57
Alcohol ethoxylate 6EO + 1% sodium chloride	51
Alcohol ethoxylate 9EO	82
Alcohol ethoxylate 9EO + 1% sodium chloride	77
Nonyl phenol ethoxylate 10EO	64
Nonyl phenol ethoxylate 10EO + 1% sodium chloride	61

Nonyl and octyl phenol ethoxylates are most versatile among the non-ionics and have held a position of prominence. However, they are based entirely upon chemicals from petroleum and are slow to lose their surfactant properties during biodegradation. Also there is some concern over suggestions that these molecules can break down to compounds that mimic hormones. Thus, the popularity of these alkyl phenol ethoxylates is waning and other ethoxylates such as the alcohol ethoxylates from synthetic or nature-made fatty alcohols are taking over.

Examples of natural fatty alcohols used for making ethoxylates are lauryl, cetyl, oleoyl and stearyl; these give lipophiles with carbon chain lengths between C_{12} and C_{18}. Coconut fatty alcohol can even provide some C_8 lipophile. Synthetic fatty alcohols offer greater flexibility in choice of lipophile but, in general, they are of about the same carbon chain lengths as the natural ones. The natural fatty alcohols all have even numbers of carbon atoms in the chain but the synthetics are not confined to this. For personal care products the natural ones are the preferred option.

Nonyl phenol ethoxylates and alcohol ethoxylates are manufactured with a range of ethylene oxide contents to provide for different functions. A rough guide is given in Table 2.3.

Table 2.3 *The degree of ethoxylation, usually quoted as moles of ethylene oxide per mole of lipophile, determines hydrophilic capacities. This, in turn, affects solubility and application; with a small ethylene oxide content the molecule preferentially dissolves in oily substances; with a large ethylene oxide content the molecule becomes water soluble*

Moles ethylene oxide	Function
1–3	Water-in-oil emulsifier
4–6	Oil-in-water emulsifier
7–12	Detergent
12 upwards	Hydrotrope, solubilizer

One type of non-ionic that is totally synthetic is made from ethylene oxide (EO) and propylene oxide (PO) by polymerizing them into the EO/PO copolymers. The lipophile is represented by repeated PO units and the hydrophile by repeated EO units as shown in Figure 2.14(c). Random and block copolymers, each having different degrees of surfactancy, are formed. Thus, at one extreme where the structure is alternating EO/PO units the resulting molecule has no surfactancy – the lipophile cancels out the hydrophile – but where separate long EO chains and PO chains exist the molecule is a surfactant.

Sorbitan esters are in complete contrast to the EO/PO polymers in that both the lipophile and the hydrophile can be plant derived. Although these esters offer little in the way of wetting and detergency properties they are excellent emulsifiers and are widely used in cosmetic preparations. An example of a typical sorbitan esters is shown in Figure 2.14(d). In addition to the straightforward esters there are the ethoxylated sorbitan esters which add to the versatility of these emulsifiers.

The following list gives some of the more common non-ionic surfactants. As with previous lists it is not meant to imply that a particular surfactant has one specific application.

Non-ionic surfactant	*Applications*
Alkyl phenol ethoxylates	
Nonyl phenol ethoxylate (9EO)	Household/industrial cleaners
Nonyl phenol ethoxylate (2EO)	Petroleum/oil emulsifier
Octyl phenol ethoxylate (10EO)	Detergents
Alcohol ethoxylates	
C_{12}/C_{14} synthetic ethoxylate (8EO)	Hand dish-wash liquids
Stearyl alcohol ethoxylate (7EO)	Cosmetics emulsifier
Cetostearyl alcohol ethoxylate (20EO)	Emulsifier, ointments/creams
Amine ethoxylate	
Coconut fatty amine ethoxylate (10EO)	Polishes
Ester ethoxylate	
Sorbitan monolaurate ethoxylate	Cosmetic emulsifier
EO/PO block polymers	
80%PO/20%EO	Low foam machine wash liquids
Alkanolamides	
Coconut diethanolamide	Shampoo foam booster
Esters	
Sorbitan monolaurate	Cosmetic emulsions
Sorbitan monolaurate 4EO	Cosmetic emulsions
Di-isopropyl adipate	Lipsticks
Cetostearyl stearate	Make-up

Amphoteric Surfactants

More recent developments in surfactants have led to the amphoterics which have a positive and a negative change within the same molecule. These surfactants offer good detergency coupled with high foaming capacity and mild action on the skin. The latter makes them particularly valuable in products such as baby shampoo, shower gel and frequent use shampoos where they act as the secondary surfactant, the primary one generally being an anionic such as sodium lauryl ether sulfate. Other example applications are listed below.

The chemical properties of amphoterics are somewhat dependent upon the pH of the water in which they are dissolved. Water with a low

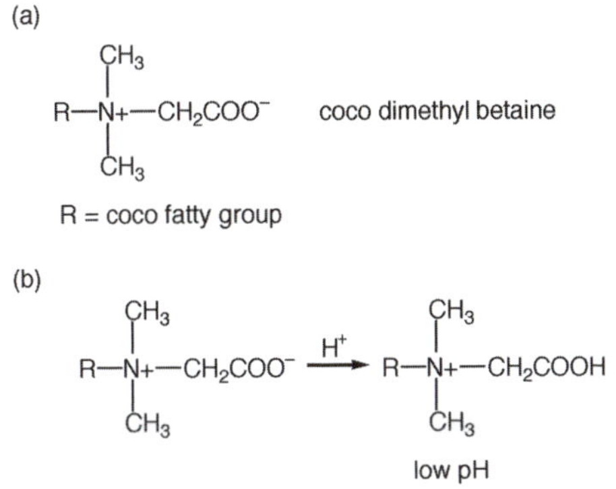

(a)

$$CH_3$$
$$|$$
$$R—N+—CH_2COO^-$$ coco dimethyl betaine
$$|$$
$$CH_3$$

R = coco fatty group

(b)

$$CH_3 \qquad\qquad CH_3$$
$$| \qquad\qquad\qquad |$$
$$R—N+—CH_2COO^- \xrightarrow{H^+} R—N+—CH_2COOH$$
$$| \qquad\qquad\qquad |$$
$$CH_3 \qquad\qquad CH_3$$

low pH

Figure 2.15 *Amphoteric surfactants. (a) Betaines, such as the one shown here, are some of the most common amphoterics. (b) Amphoteric surfactants are pH sensitive*

pH, acidic, results in an amphoteric molecule behaving as a cationic molecule, as its negative charge attracts positive hydrogen ions from the acidic solution as outlined in Figure 2.15.

Amphoteric surfactant	*Applications*
Coco imidazoline betaine	Cold water detergent
Coco amido sulfo betaine	Acid metal cleaners
Oleo amido propyl betaine	Skin care products
Tall oil imidazoline	Industrial detergent

Surfactants, the Future

In the main, surfactants are made from petrochemicals by processes such as sulfonation, ethoxylation and hydrocarbon modification *etc.* However, a growing number of surfactants are now manufactured from oleo-chemicals, plant-derived chemicals. For example, plant alcohol ethoxy-lates in which the lipophile is of vegetable origin are now widely used in cosmetics but ethoxylation to produce the hydrophile still relies upon petroleum.

Most notable in terms of the environment/resource equation are the alkyl glucosides and sucrose esters (Figure 2.16). These are environ-mentally benign in that both parts, the alkyl part and the carbohydrate part, are plant derived rather than petroleum based. The alkyl group is typically a plant alcohol such as lauryl alcohol, $C_{12}H_{25}OH$, obtained from hydrolysis of coconut oil or palm kernel oil. Where glucose is used as the basis of the hydrophile a whole range of materials – some of them waste products from other processes – is available because of the ubiquitous nature of glucose-based compounds in the plant world.

Both lipophile and hydrophile retain most of their natural structure and there are no nasties such as ethylene oxide as essential requirements. Alkyl glucosides are thus from renewable feedstocks and are readily bio-degraded once in the environment. As such they appear to have a particu-larly healthy 'cradle-to-grave' profile but applications progress has been slow due to high cost. They are, however, making inroads into food, cosmetics and industrial degreasing preparations. Although they are resistant to high levels of alkalinity they are unstable towards acids and tend to break down.

This approach to building surfactant molecules appears to be the way things may move in the future to satisfy demands for chemicals made from renewable resources (plants) and lessen the reliance upon petroleum feedstocks.

Of all the surfactant molecules looked at so far the essential structure has been a single long chain lipophile attached to a single hydrophilic head. We have seen how this gives rise to certain properties and how

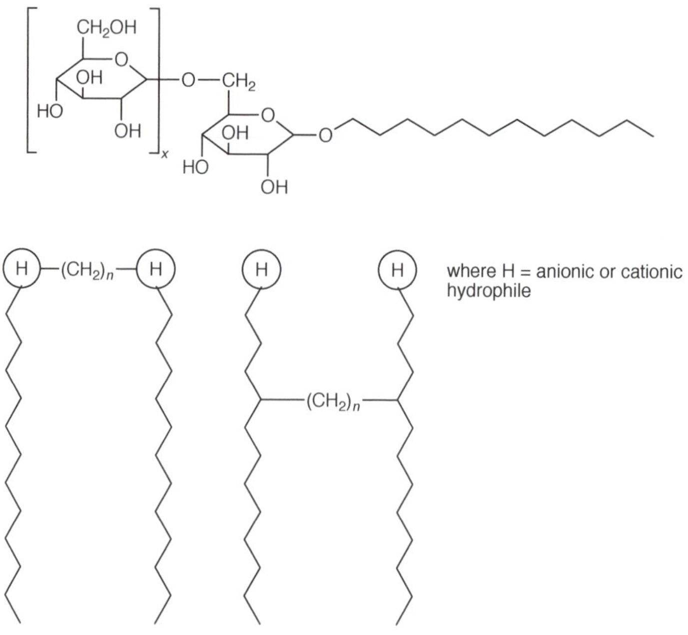

Figure 2.16 *In alkyl glucosides both the alkyl group (lipophile) and the glucose (hydro-phile) come from plant sources. Gemini surfactants or dimeric surfactants have a twin-like structure and represent a new and exciting development*

simple modifications, such as increasing chain length or putting in ethylene oxide groups to increase the size of the hydrophile, can bring about the properties required.

A more radical approach to the structure of the surfactant molecule is to have a single molecule with two lipophiles and two hydrophiles. This is exactly what is done in a new breed of surfactants called Gemini surfactants or dimeric surfactants.

Thus, two fairly conventional looking surfactant molecules are twinned by means of a linking group attached either to their hydrophilic heads or to the hydrophile ends of the lipophiles. A particular example here is the acetylenic diol, a non-ionic Gemini, in which the linking group is carbon–carbon triple bond. This bond, the acetylenic bond, is a centre of high electron density and adds to the polar nature of the hydrophiles.

When adsorbed at the water/air interface these molecules have their lipophiles lying flat on the surface unlike regular surfactant molecules that have their lipophiles projecting away from the surface. This arrangement hinders foam formation and makes the acetylenic diols the preferred surfactants where foam would be detrimental such as in surface coatings.

There is much current research in new Gemini molecules as the structures, whether non-ionic or ionic, provide high levels of surface activity. One area in which Geminis show promise is as vehicles for delivering bioactive molecules into living cells.

HYDROTROPES AND VISCOSITY

Many formulations, especially those used in heavy degreasing/cleaning operations as is required in some industrial processes are made in concentrated form to minimize space demands in storage and transport, reduce packaging waste and to offer one product that may be diluted to different strengths for different demands. Among concentrated solutions it is common to find formulations that have high electrolyte contents (phosphates, silicates, caustics *etc.*) as well as maximized levels of surfactants.

Preparing such formulations poses the problem of how to put the whole together and obtain a stable solution and keep the water to the minimum. Much of the difficulty in doing this relates to the fact that the concentration of surfactant is greatly in excess of the CMC and the micelle structure of the surfactant is going to be that of the high viscosity liquid crystal phases. And the large amount of electrolyte makes matters even worse.

To overcome this hydrotropes are used. In hydrotropes the lipophilic chain is relatively small compared with the hydrophilic head unlike surfactants where the opposite applies. This structure enables the hydrotrope molecules to aggregate with the surfactant molecules and become a part of the micelle structure. The large hydrophilic heads relative to the small lipophilic tails gives them a wedge-like shape as shown in Figure 2.17, which enables them to force their way between the surfactant molecules in the micelle and destabilize the structure. As a result the formation of large micelles is prevented and the viscosity kept to a low value. Examples include: sodium toluene sulfonate, sodium cumene sulfonate, sodium xylene sulfonate. These are the mainstay hydrotropes but new ones such some of the alkyl glucosides and amphoterics are starting to play a role here. Most hydrotropes are poor in terms of surfactancy but a few of the surfactants themselves can behave as hydrotropes.

Knowing how much and what type of hydrotrope to use in different situations is not easy and tends to be an empirical exercise. The use of a hydrotrope selection guide, from surfactant suppliers, removes much of the trial and error aspect for formulations based on common surfactants in combination with alkaline builders.

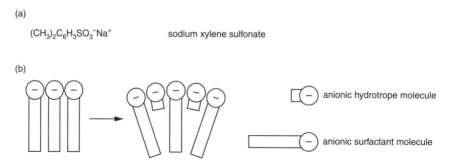

(a)

$(CH_3)_2C_6H_3SO_3^-Na^+$ sodium xylene sulfonate

(b)

Figure 2.17 (a) *A commonly used anionic hydrotrope, sodium xylene sulfonate, has a large hydrophilic head group and relatively small lipophilic tail.* (b) *Hydrotrope molecules become wedged between the surfactant molecules of the micelle, thereby destabilizing the structure and preventing growth to larger structures which would increase the viscosity*

EMULSIFICATION

Of all the different types of everyday formulations it is those that involve the preparation of emulsions that pose the greatest number of problems. The reason: emulsions are energetically unstable systems and as such are constantly under influences trying to make them revert to their component parts. Some emulsions will do this within only seconds of being formed whereas others have a sufficient degree of stability to give them a

life (usually reported as half life) of several years. Another way of looking at it is to see emulsions as being energetically (thermodynamically)
unstable but kinetically metastable systems.

Emulsions are dispersions of two immiscible liquid phases, usually oil
and water, and can be of two different forms: oil-in-water (o/w) emulsion
or water-in-oil (w/o) emulsion. In o/w emulsions the oil is dispersed (the
disperse phase) as small particles within water (the continuous phase)
and for w/o emulsions the exact opposite is the case – see Figure 2.18.

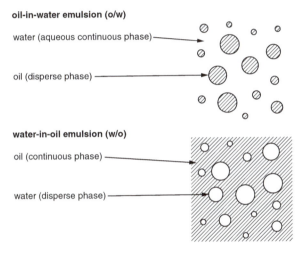

Figure 2.18 *Emulsion types:* (a) *Oil droplets dispersed in aqueous continuous phase, oil-in-*
water, o/w emulsion; (b) *Water droplets dispersed in oil as the continuous*
phase, water-in-oil, w/o emulsion

Surface activity is at the heart of emulsions because these systems are
dominated by interfacial phenomena. To appreciate the importance of
surface chemistry in emulsions consider the following example.

Placing 50 cm³ of oil and 50 cm³ of water into a 6 cm diameter jar
results in a layer of oil floating upon the water with a well defined oil/
water interface, the area of which is

$$A = \pi r^2 = 28.3 \text{ cm}^2$$

When only a few drops of emulsifying agent are added, the jar capped
and vigorously shaken the oil becomes dispersed as tiny droplets. If the
droplets have a typical diameter for emulsions of 0.0001 cm the volume
of each one is

$$V = 4/3 \, \pi r^3 = 523 \times 10^{-15} \text{ cm}^3$$

From this and the total volume of 50 cm^3 the number of oil particles is

$$N = 0.0956 \times 10^{15}$$

And the total surface area of these is

$$A = 4 \pi r^2 N = 3\,000\,000 \text{ cm}^2$$

Thus a one-hundred-thousand fold increase in surface area accompanies the emulsification process. Clearly the surface becomes a more significant entity the greater the degree of sub-division of the disperse phase and the unusual properties associated with surfaces become pronounced. In general, for o/w emulsions, the disperse phase (oil) amounts to between 5 and 60% but for w/o emulsions the disperse phase (water) may be as high as 70%.

An everyday example of an o/w emulsion is mayonnaise; this is essentially tiny droplets of vegetable oil, the disperse phase, in water (vinegar = 4% acetic acid in water), the continuous phase. Simply shaking together the vinegar and oil results in a temporary dispersion of oil in water approximating to a highly unstable emulsion. Within but a second or so the whole has separated to a layer of oil floating upon the aqueous layer.

To produce a reasonably stable emulsion of oil and vinegar requires the presence of an emulsifier, a surfactant, and in the mayonnaise example this comes in the form of egg yolk. Combining the correct amounts of the three ingredients with some dedication to procedural details results in an emulsion stable enough to last for the present meal and for future meals – if stored with care.

Attempts to add other ingredients to the creamy concoction may well turn out to be a demonstration of emulsion instability. Similarly, if the mayonnaise is subjected to temperature excursions this may also be sufficient cause for it to revert to the two separate layers from whence it came.

Stability of Emulsions

In addition to emulsions breaking into their two phases as a result of their inherent instability they may also undergo a process of inversion from, say, o/w to w/o. A common example of this is in the dairy where the churning of cream, an o/w emulsion is changed to butter, a w/o emulsion. In this inversion process some of the water present in the cream (as the continuous phase) is displaced as its role changes to become the disperse phase.

This displacement of one component is a common feature of the inversion process. Tests to distinguish between o/w and w/o emulsions are outlined in the analysis and testing section. Where an o/w emulsion has been formed using a non-ionic ethoxylate as emulsifying agent then inversion to the w/o form may occur on heating. This is a result of the ethoxylate chain becoming dehydrated as the thermal energy of the water molecules surrounding it is sufficient for them to break free from their hydrogen bonds – the cloud point effect. The lipophilic character of the molecule becomes greater making it more effective as a w/o emulsifier and inversion results.

The surfactant chosen – how it is chosen is considered later – for the emulsification process must, if the emulsion is to reasonably stable, produce an interface between the two phases that is highly elastic. This is needed to withstand the mechanical action in which, for example, a droplet of the disperse phase migrates about the continuous phase and collides with other dispersed droplets.

A weak interface would be ruptured by such a collision resulting in the colliding particles coalescing – the onset of separation. In a similar vein if the emulsion is prepared to have large particles of the disperse phase it will be less stable than one of similar composition but having smaller disperse phase particles.

Improvements to emulsion stability can be made by using two structurally different surfactants. Such a mixture produces an elastic adsorbed layer as a result of the different shaped lipophiles becoming tangled or interwoven and less easily separated than the orderly arrangement of lipophiles of the same shape. In reducing the probability of rupture of the interface during collision viscosity plays a vital role. Increasing viscosity reduces the impact energy and the frequency of collision; both these go towards decreasing particle rupture. Higher viscosities are generally achieved by either incorporating thickeners although careful choice of surfactants can go a long way to achieving a suitably high viscosity.

Temperature increases affect emulsion stability in various ways: changing the solubility of the emulsifying agent as outlined above; reducing the viscosity of the phases, particularly that of the continuous phase which is often critical; causing greater rate of diffusion (rate of diffusion is proportional to temperature) and agitation between the phases along with more collisions between particles.

The HLB System

The problem of selecting the most suitable surfactant for a particular emulsification task is formidable when one takes into account the huge

number of different commercial surfactants and that a pair of surfactants in a particular ratio is likely to be the best option. To set out and
do a trial and error exercise would be a mammoth task.

Fortunately, help is to hand in the form of the HLB (Hydrophile
Lipophile Balance) numbers for selecting emulsifiers. It is, for the most
part, confined to non-ionic surfactants, which give values between 0 and
20, to indicate the relative strength hydrophilic and lipophilic parts of
the surfactant molecule. HLB numbers are calculated, on the basis of the
surfactant's molecular structure, by the manufacturer and are provided
along with the other technical data.

Although HLB is primarily a calculated figure it is possible to obtain
an HLB by comparing the unknown with a surfactant or blend of known
HLB in emulsifying a chosen oil. By such means it is possible to obtain
HLB values of anionic surfactants, quantities that cannot be obtained by
calculation.

Molecules with low HLB values, 0 to 10 (weak hydrophile, strong
lipophile), are oil soluble and are useful as w/o emulsifiers. The ones with
high HLBs, 10 to 20, are water soluble and can be used as o/w emulsifiers.
Most of the HLB data available is for the preparation of o/w emulsions
but the system may also be applied to w/o emulsions. A few ionic surfactants have been assigned HLB values, for instance: the soaps, sodium
oleate (HLB 18), potassium oleate (HLB 20), sodium lauryl sulfate (HLB
greater than 20).

Although the greatest use of HLB values is in emulsion work the
values can be used in other applications since there is an optimum
HLB value for best performance in wetting and detergency. The first
task in preparing a stable emulsion is to find a surfactant system that
has an HLB number of the same value as that of the proposed oil
water blend.

Consider for example making an o/w emulsion. The starting point is to
find the HLB requirement of the oil phase and then to match this with a
surfactant or mixture of surfactants which give the same HLB value. If
the oil phase is simply made up of a single substance then it is a straightforward matter of looking up the HLB for that substance. However, as is
usually the case, the oily phase comprises a mixture of substances. As
such the HLB has to be calculated. Tables giving HLB requirements for
different oils (and fats, waxes, solvents *etc.*) to make oil in water emulsions are available, as are tables listing the HLB numbers for surfactants
from surfactant manufacturers.

HLB values for surfactants cannot be directly measured as one would,
for example, measure the density of a liquid. However, an approximate
indication of HLB range may be obtained by observing how the

surfactant behaves when mixed with water followed by thorough agitation as shown below.

Does not disperse and readily separates	HLB 1 to 4
Disperses with difficulty	HLB 3 to 6
Forms a milk	HLB 6 to 10
Cloudy to clear dispersion	HLB 10 to 13
Solution	HLB 13 to 20

HLB values give an indication of the main function as shown below. These would typically be found with nonyl phenol ethoxylates.

Water-in-oil emulsifier (1–3 EO)	HLB 3 to 6
Wetting agent	HLB 7 to 9
Oil-in-water emulsifier (4–6 EO)	HLB 8 to 15
Detergent (7–12 EO)	HLB 12 to 15
Hydrotrope (12 & over EO)	HLB 15 to 18

In the preparation of emulsions it is common practice to heat the oils/waxes until a fairly thin homogenous melt forms. Surfactants are added to either or both phases according to their respective solubilities before the two phases are blended together. Thus, low HLB emulsifiers are oil soluble and so they are added to the oil whereas the higher HLB members, the water soluble ones, are dissolved in the aqueous phase.

Mixing typically requires a temperature of about 60 °C but for high melting point waxes such as Carnauba 90 °C is needed to get a thin melt suitable for mixing. The aqueous phase must also be heated to a similar temperature so as not to freeze out the wax upon combining the two liquids. Blending is typically carried out by adding, bit by bit, the disperse phase to the continuous phase with thorough mixing, sometimes with high shear stirrer, and then allowing to cool whilst stirring continues.

Example

Consider the following preparation for a skin cream based on an o/w emulsion in which an oil and two waxes are emulsified in water. Because this is an o/w emulsion the continuous phase is water and this will represent the major proportion, typically about three-quarters of it, leaving the oil phase to account for about a quarter.

Assume it has been decided that the oils required and their proportions lead to an oil phase comprising 80% medium liquid paraffin, 12%

cetostearyl alcohol, 8% lanolin. A look through HLB tables gives the values and enables the HLB requirement to be calculated as follows.

Liquid paraffin, medium mineral oil, 80%	HLB 10, 80% of 10 = 8
Cetostearyl alcohol, 12%	HLB 15, 12% of 15 = 1.8
Lanolin, 8%	HLB 9, 8% of 9 = 0.72
HLB requirement	*= 10.52*

In general we would work either side of this figure, say, 9.5 to 11.5. The chemical nature of the surfactants needs also to be taken into consideration. For this system surfactants of the sorbitan monostearate type are suitable – surfactant suppliers can often recommend the usual surfactant type for a particular job but it is up to the formulator to work out what the HLB requirement for his formulation.

The HLB requirement simply represents a starting point. Of course a single surfactant may well provide the exact HLB but it is often better to use a blend of two surfactants with a good spread of HLB values as shown below. Looking at the HLB values of the two surfactants suggested here it is seen that ethoxylated one is weighted on the hydrophilic side (water soluble) whereas the other surfactant is towards the lipophilic side (oil soluble).

Sorbitan monostearate (20EO)	HLB 14.9
Sorbitan monostearate	HLB 4.3
Mixture 1:1 of above	HLB 9.8

The amount of surfactant mixture to use is normally around 2 to 5% based on the whole formulation where the oil phase is about 30% and the aqueous phase (containing surfactant) about 70%.

SOLUBILIZATION

Another way in which surfactant molecules in aqueous solution can emulsify an oil is by solubilization. This results in liquids, called microemulsions, in which the oil is held within the surfactant micelle structure as shown in Figure 2.19. Normally the spherical micelles have a diameter of about twice the length of the lipophile and this is capable of some expansion as other lipophilic material, *e.g.*, an oil, dissolves within the micelles.

Although the micelle is then swollen it is still colloidal. With particles of dimensions 0.01 to 0.1 μm, too small to scatter light, these microemulsions are transparent, or have only a slight bluish opalescence, unlike regular emulsions (macro-emulsions) in which the particle sizes are in the region of 0.1 to 10 μm and large enough for the light to be scattered producing the characteristic white opacity.

Non-ionic surfactants, particularly the ones with high HLB values, are more effective than ionic surfactants at solubilization. Also, surfactants with the ability to produce larger micelles are better at solubilization.

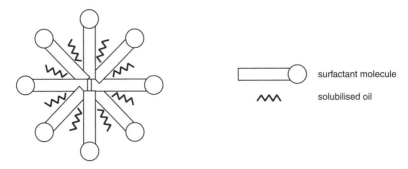

Figure 2.19   *Micro-emulsions are made up of surfactant micelles in which the oil is solubilized in the micelle interior. These are sometimes thought of as swollen micelles*

Essential oils such as eucalyptus exemplify the micro-emulsion and macro-emulsion particularly well. If the oil, which is immiscible in water, is mixed with about three times its mass of a non-ionic ethoxylate and then added to water the result is a clear solution, or to be more precise, a colloidal dispersion of swollen micelles. On the other hand if the oil and water are mixed in the presence of only a small amount of the same ethoxylate a milky emulsion is formed. We could therefore make a preparation of eucalyptus oil as a white milk or as a clear solution.

FOAMING

A foam is a dispersion of a gas in liquid, for example shaving foam, or a gas in a solid as in foam rubber. Usually the liquid is an aqueous solution and the gas is air but for aerosol dispensed foams the dispersed gas is the propellant such as propane/butane. Foams have unique and very useful properties that come about, as with emulsions, as a result of the huge amount of interfacial area; in emulsions the interface is oil/water whereas in foams it is water/gas.

Thus the study of foams is in some respects similar to that of emulsions and involves an understanding of the adsorption of surfactants at the water/gas interface. To appreciate the importance of the surface in foams, a calculation similar to that for emulsions could be carried out.

Different foam structures can be formed depending upon the proportion of dispersed gas to liquid. In Figure 2.20 some extreme forms are shown.

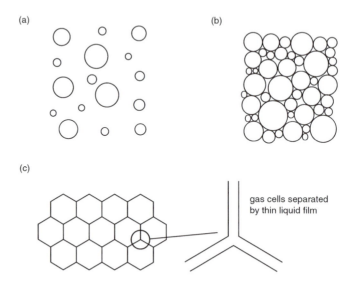

Figure 2.20 *Different forms of foam. (a) Gas dispersed in large volume of liquid. (b) With less liquid bubbles start to pack together. (c) A minimum of liquid results in polyhedral cells of gas in which liquid drains from between the intercellular surfaces. The mechanism of drainage depends upon the surfactant and may lead to rupture of the film and foam collapse (as with anti-foams) or to a stable structure capable of withstanding mechanical stress due to high flexibility (as with foaming agents)*

In the first example it will be seen that the liquid will tend to drain downwards under the influence of gravity, or, looking at it another way the bubbles will tend to float upwards. The result will be that the amount of liquid held in the foam decreases and the bubbles move closer together until polyhedral gas cells shown in part (c) of Figure 2.20 are formed. Drainage of the liquid may well proceed no further and a relatively stable foam results. However, if the liquid continues to drain the thin film will rupture and the foam collapse.

For a stable foam to be produced the thin aqueous film that forms the cell walls needs to have the correct surface properties, namely elasticity and capacity to resist the continuous drainage that would, by gravity or capillary action, cause excessive thinning to the point of rupture. With the right surfactant these surface properties can be achieved.

Foam is elastic and can remain stable despite being pulled about and this is due to the Gibbs–Marangoni elasticity effect. As the film stretches under an applied stress the area increases and amount of surfactant adsorbed at the two gas/liquid surfaces of the cell walls falls. This causes an increase in surface tension and an elastic effect. Surfactant molecules from within the cellular liquid then diffuse to the two gas/liquid surfaces to make up the shortfall and restore the surface tension to its former value. This diffusion takes time during which a relatively high local surface tension exists to cause the stretched film to recover its original thickness.

DETERGENCY

By far the largest volume consumption of surfactants is in detergent products. The way in which detergents work is a complex topic because it involves, to different extents, many different functions. It is further complicated by the variety of substrates, and mixtures of different types of soiling. For a surfactant system to succeed in detergency it must first of all be effective at reducing the interfacial tension between water and soil particles and water and substrate materials so that wetting of the various surfaces results.

In the examples of surfactants that have been considered in previous sections it is the anionics that have become pre-eminent in the field of detergency. Thus, sodium alkylbenzene sulfonate and sodium lauryl sulfate are popular for washing-up liquids and machine clothes wash preparations. These anionics are usually the primary surfactants and blended with other surfactants such as amphoterics and non-ionics. The number of soils that need to be removed by a detergent is large and some grouping is necessary for an understanding of the process of dirt removal from clothes, crockery and bodies.

Most soils encountered in detergency considerations comprise a mixture of dust particles such as finely divided minerals (silica, brick, coal, cement, soot, metals) and organic substances (smoke, fat, grease, protein, food). Nearly all these are acidic materials or contain chemical groups that yield acidic substances during hydrolysis. As such many soils tend to react with alkalis and, in the process, form water soluble salts if the right alkali is chosen.

Thus the pH during washing plays an important role in detergency and a first option in soil removal is to go for alkaline wash solutions. In fact, for many soils, a good degree of cleaning can be achieved by alkali itself with no detergent present. Most of the fats and greases of interest melt at around 40 °C so if the wash water is hot this enhances the removal of oily soiling. Thus, alkali and hot water can go a long way to removal of soil. A traditional example is the use of a hot solution of washing soda to soak the dirty laundry in.

High alkalinity is something that may be exploited in automated cleaning systems such as ones used in industrial processes and where there is no personal contact with the chemicals. It is, however, too severe for most household uses, especially those that come into contact with the skin. Caustic oven cleaners, containing sodium hydroxide, for removal of heavy baked on grease, are perhaps an exception here but these have to be used with a considerable degree of caution.

Household products can utilize alkalinity only to a very limited extent and in some examples the exact opposite, low pH, is opted for. Washing machine preparations and machine dishwash formulations can take advantage of fairly high pH and high temperatures.

Alkaline conditions also suppress the solubility of water hardness salts such as the calcium ions which are detrimental to the washing process due to a number of interferences: reacting with soaps to form scum, reacting with anionics to bridge the hydrophilic heads and reduce their effectiveness as surfactants, reacting with soils to form insoluble compounds *etc*. In addition to alkaline conditions the problem of calcium ions and the like can be tackled by addition of chelating agents.

The detergent effect of surfactants comes about when surfactant molecules present themselves for adsorption at the soil/water interface, soil/substrate interface and substrate/water interface. Molecules adsorbed at the water/air interface are not directly involved in detergency but account for foam formation and so act as an indicator to show when the detergent has been used up. The rate at which a surfactant molecule arrives from the bulk liquid at these interfaces depends upon its concentration in the solution as the monomer, the unassociated molecule. Surfactants that are able to produce high concentrations of monomer, those that have high CMC values, are particularly effective in this respect, which explains the importance of the anionics in detergency.

In a typical case a proportion of the soil that is not bound by the oily substances will have been dissolved, assisted by the alkali, and migrated from the substrate surface to the wash solution. Some of the oily soil will contain fats (triglycerides) and fatty acids that react with alkali to form the fatty acid salts which become dispersed in the water. These fatty acid salts are, in fact, soaps and so we have a saponification reaction occurring during the washing process. Remaining oily deposits will have been softened by the hot water and, where these are thick deposits, elevated temperature may well reduce the viscosity to such a level that gravity alone is sufficient and the oil simply floats off. Remaining oily material requires surfactant molecules for its removal.

Surfactant molecules arriving at the oily soil become adsorbed with their lipophilic tails sticking into the oil and their hydrophilic heads to the outside where they are attracted to water molecules as shown in Figure 2.21.

Where micelles of surfactant come into contact with the oily soil there arises the possibility of the oil being solubilized into the lipophilic core of the micelle as was seen in the section on micro-emulsions. This effect is, normally, only a minor contribution to detergency as most of the work

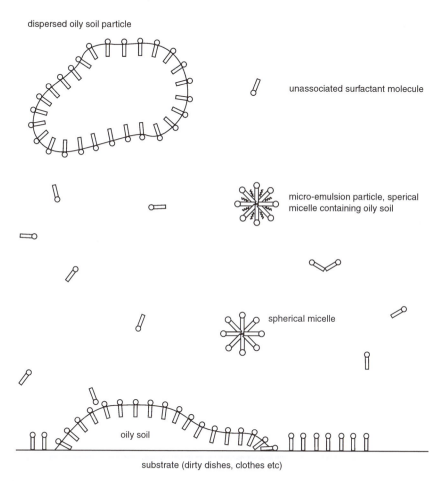

dispersed oily soil particle

unassociated surfactant molecule

micro-emulsion particle, sperical micelle containing oily soil

spherical micelle

oily soil

substrate (dirty dishes, clothes etc)

Figure 2.21 *Surfactants in detergency. Oily soil is removed from the substrate by several actions involving surfactant molecules in water. The lipophilic tails of the molecules are attracted to the oily soil and adsorb onto it with their hydrophilic head groups oriented towards the water. The oily soil is thus dispersed into the water in a similar manner to the oil-in-water emulsion formation. Simultaneously, surfactant molecules adsorb onto the substrate surface with their hydrophilic heads towards the water, preventing the oily soil being re-deposited. Where the concentration of surfactant is sufficiently high for it to exist as micelles a proportion of the oily soil may be removed by solubilization to form a micro-emulsion*

is done by the monomer. However, as the surfactant concentration increases above the CMC the role played by micelles is more significant.

It is at this point worth emphasizing the dynamic nature of the system. On a molecular level the whole of the liquid consists of molecules and micelles in rapid random motion. Unassociated surfactant molecules are constantly arriving to adsorb at the various interfaces whilst simultaneously others desorb from the interface and diffuse back into the bulk of the liquid.

The micelles themselves have molecules entering and leaving their structures and some are collapsing whilst at the same time others are being built up. Thus a process like solubilization is not to be thought of as hydrophobic molecules squeezing into an established structure rather it is one of hydrophobic molecules being attracted to the hydrophobic surfactant chains at the moment they are associating to form a micelle.

Adsorption of the surfactant molecule onto the substrate surface is also important. When the strength of adsorption between a surfactant lipophile and substrate surface is greater than that from the oil-to-substrate adsorption then surfactant molecules move towards the oil, at the same time wetting the substrate surface, and forcing the oil away in a kind of roll-back manner.

Chapter 3

Formulations

A few formulations are considered here to see how the various chemicals that were studied in the previous sections function when combined in a product. Of all the chemicals we find in modern everyday formulations surfactants are by far the most widely used because of their amazing versatility.

One familiar group of products is the personal care range and this is taken as the starting point. Our familiarity is not confined to our intimate contact with personal care preparations but extends to the frequency with which we use them. A typical shampoo is considered here and we are given a head start in the form of a list of ingredients on the back label of the bottle.

In the formulations studied here and in subsequent sections amounts are given as percentage weight/weight (%w/w) as is now common.

SHAMPOO

The following details were taken from the label of a supermarket brand of shampoo. On its label is a list of the ingredients in terms of the CTFA or INCI names as is required for personal care products. The ingredients shown below make up what is a fairly complex chemical system and to understand the chemistry it is necessary to consider the ingredients and how they interact.

Ingredient	*Chemical names*
Aqua	Water
Sodium laureth sulfate	Sodium lauryl ether sulfate, anionic surfactant
Sodium chloride	Sodium chloride, common salt
Cocoamide DEA	Coconut diethanolamide, non-ionic surfactant
Coco amido propyl betaine	Coconut amphoteric surfactant

Parfum	Fragrance
Citric acid	Citric acid (2-hydroxypropane-1,2,3-tricar-boxylic acid)
Polyquaternium	Quaternary cationic protein derivative
Formaldehyde	Formaldehyde (methanal)
Tetrasodium EDTA	Tetrasodium ethylenediaminetetraacetate
Methyl paraben	Methyl 4-hydroxybenzoate
Propyl paraben	Propyl 4-hydroxybenzoate
CI 47005 & 74160	Colour Index numbers, Quinoline Yellow and Copper Phthalocyanine

Although the manufacturer has listed the ingredients the actual amounts are not reported. It was seen in previous sections that chemicals, particularly surfactants, may interact in a synergistic manner and that this is at the heart of formulation work where the aim is to get maximum performance with the minimum of ingredients. Knowledge of the quantities used is vital information and manufacturers are, quite naturally, unlikely to reveal it. However, they are required to list the ingredients in order of the amount present starting with the one in greatest proportion.

With a basic knowledge of surfactants and related chemicals we can deduce what the function of each ingredient is and, with a little experience in formulation work, we can also suggest the likely amounts of those ingredients. In the following sections each ingredient is discussed in turn.

Water

It may come as a surprise to see that aqua (water) comes top of the list but, as will be seen later, a large amount of water is necessary to get the correct properties for the product. Of course, manufacturers of shampoos do not complain about the fact that what they are selling is largely water. But the water is generally not just tap water; it will usually be filtered and deionized.

Sodium Lauryl Ether Sulfate

The overall function of a shampoo is to create a sense of well-being in the user by turning their dirty hair into clean hair. The main shampoo ingredient to help achieve this is the primary surfactant. In this formulation sodium lauryl ether sulfate, probably with 2 or 3 moles of ethylene oxide, carries out the biggest part of the cleaning action and that is to remove, from the hair shaft, the natural greases such as sebum that have built up and which bind particles of dust and dirt. Once this grease has been transferred into the water along with the released dirt particles the essential function is complete.

Removing grease and dirt, could be partly accomplished by water itself, especially if the water was hot (for comfort not greater than 40 °C) but for a thorough cleaning action a surfactant is required. Sodium lauryl ether sulfate (3EO), an anionic surfactant, is commonly chosen for shampoo formulations.

All shampoos are, in effect, fairly strong solutions that become diluted at the point of use. To enable proper dilution the concentration of the sodium lauryl ether sulfate is important – too strong and it will gel when added to water, too weak and several squirts from the bottle will be required much to the annoyance of the user. For convenience sodium lauryl ether sulfate (3EO) is supplied as a diluted liquid such as a 28% solution – referred to as 28% active matter.

Typically this surfactant would be used at 40% in a shampoo giving an active matter content of 11.2%. At this concentration in pure water, a clear, fairly thin, liquid would be expected. In this solution the surfactant would be well above the CMC level and we can assume that most of the surfactant molecules are associated in the form of spherical micelles.

A simple surfactant solution like this would be perfectly adequate at washing the hair – that is, at degreasing the hair. However, when poured from the bottle or applied to the hair most will be wasted as it dribbles between the fingers and rapidly drips off the hair and into the wash basin. Clearly a wishy-washy liquid is inadequate as a shampoo and it needs to be thicker.

Having a greater concentration of sodium lauryl ether sulfate would achieve this but when highly concentrated surfactant solutions come into contact with water, especially tap water which contains electrolytes, a rigid gel may form which is difficult to re-dissolve.

To understand more on viscosity effects of surfactants leads us a consider of micelle structure and how that explains our observations. To get greater viscosity it is necessary to increase the number of the larger micelles such as the cylindrical ones and the hexagonal arrays of them.

Unassociated surfactant molecules do nothing for viscosity, spherical micelles are little better but the large cylindrical structures and the hexagonally packed structures have a large viscosity effect. For the shampoo it is therefore desirable to arrange for the formulation to have a high proportion of surfactant in these cylindrical forms.

Sodium Chloride

The effect of an electrolyte in a sodium lauryl ether sulfate solution is to promote sphere-to-rod transitions and cause a marked increase in viscosity as there is then much more contact between large elongated structures than is possible with small spherical micelles.

Careful control of the amount of salt added will result in the right amount of viscosity increase giving the liquid greater staying power and stop it heading straight for the plughole. Only small amounts of sodium chloride are used as at higher concentrations the electrolyte effect causes the surfactant molecules to pack into the lamella structure which produces low viscosity as the layers easily slip under an applied force. Sodium chloride is typically used at around 3%.

It may be desirable to have even higher viscosity but pushing the electrolyte effect to its limits has its accompanying problems such as clouding, when stored at low temperatures, the resulting blend may also take on a slimy texture. Other contributions to thickness need to be considered and this partly explains the next ingredient on the list.

Coconut Diethanolamide

This is a non-ionic surfactant whose main function is to improve foam stability by changing the structure from one of coarse bubbles to very fine ones. Coconut diethanolamide simultaneously increases viscosity by promoting even more of the larger micelle arrangements. The overall effect of having this non-ionic present is to boost foam, and provide a thick and creamy texture.

It may well be that foam is purely a psychological factor when it comes to, say, hand dish-wash detergents but for shampoos foam also offers technical benefits. It helps to build the shampoo around the mass of hair, separating the strands. Furthermore, flexibility of the foam enables it to constantly reform itself and assist retention of the available surfactant. In shampoos it is usual to add the coconut diethanolamide at the 1 to 2% level. Putting together the few ingredients considered so far would give us a basic shampoo formulation but the commercial product from the supermarket still has more chemicals.

Coco Amido Propyl Betaine

This is an amphoteric surfactant capable of improving the degreasing action without an increase in irritancy. As such it would be regarded, in this formulation, as the secondary surfactant. It is worth noting that, although the betaine plays a small role in this shampoo, it may well be much more significant in shampoos where a high degree of mildness and very low irritancy are demanded such as in baby shampoos. In the formulation under consideration here the amount of coco amido propyl betaine is likely to be around 2%.

The gain in popularity of the betaines is due to the combination of good detergency with mildness and low irritancy to skin and eyes. By cautious use of a secondary surfactant such as this one it is possible to boost surfactancy without, at the same time, producing a harsh product that would leave the hair squeaky clean. Being squeaky clean might be acceptable for washing dishes but not for washing hair. It is likely that the formulator of my supermarket shampoo would have carried out synergy studies to optimize the performance of the amphoteric/anionic blend.

Fragrance

Small amounts of fragrance are added to mask the smell of the chemicals and provide an odour that is pleasant and fulfils the users expectations. It is likely that this is a fragrant oil, probably a blend of synthetic esters, aldehyde, ketones, terpenes *etc.*, and is present as an emulsion.

Citric Acid

A natural tribasic acid, 2-hydroxypropane-1,2,3-tricarboxylic acid, in which the first K_a of 7.45×10^{-4} is large enough to bring about useful acidity but is not so large that a slight excess slight excess of the acid would make the product hazardous. Strong acids, such as the mineral acids, with their large K_a values (orders of magnitude stronger than citric) have no role to play here – a shampoo declaring on its label of ingredients that it contains sulfuric acid would not attract much enthusiasm in the market place.

In the shampoo, citric acid is used for pH adjustment (most likely to between 5 and 6) giving a slightly acidic product that minimizes skin irritancy. In place of citric acid another natural acid, lactic acid ($K_a = 1 \times 10^{-4}$), is popular as it helps make the skin soft and smooth.

Quaternary Cationic

As noted when considering the different types of surfactants a cationic would, in general, react by means of ionic bond formation with an anionic surfactant molecule to produce a large hydrophobic molecule with no surfactant activity. However, some quaternized protein molecules are compatible with the likes of sodium lauryl ether sulfate and can be used as conditioners. They bond to the hair shaft and give a molecular coating that leaves the hair with shine, a silky feel and imparts it with anti-static properties to prevent fly-away hair during drying and combing. Thus the hair is conditioned and manageable even when only small amounts of the cationic are present at concentrations of 1 to 2%.

Tetrasodium EDTA

This chelating agent, tetrasodium ethylenediaminetetracetate, is present to sequester metal ions and prevent them from having adverse reactions with the other ingredients. For example, during the manufacture of the shampoo it would inevitably acquire traces of metals (iron, copper *etc.*) from the raw materials and the equipment in the manufacturing processes.

Metals such as these can have a catalytic effect upon the shampoo ingredients, bringing about undesirable reactions and, because the effect is catalytic, the participating metals are not consumed and can continue their destructive effects for a long time. In addition there will be the metal ions from the tap water used in washing the hair. These ions are not a problem due to any catalytic action as the time they are mixed with the shampoo is small but some of the ions, particularly calcium ions from the lime in the water, can instantly have an adverse effect on the anionic surfactant by coupling the anionic head groups and impeding surfactancy.

Although the pH conditions may result in the EDTA not working at greatest efficiency its presence in shampoos and related products is widely accepted as an effective sequestrant. Normally only small amounts, about 0.1%, are needed.

Formaldehyde, Methyl Paraben and Propyl Paraben

Formaldehyde (methanal) is a gas at room temperature. Because of this it is obtained and used as an aqueous solution such as Formalin which is 37%w/w formaldehyde. Currently there is concern over the health and safety aspects of this as a preservative and its presence in personal care products may not have much of a future. However, it is cheap and effective at killing a wide range of the micro-organisms that can cause spoilage.

The presence of the methyl paraben (methyl 4-hydroxybenzoate) and propyl paraben (propyl 4-hydroxybenzoate) broadens the spectrum of micro-organisms that can be killed. Preservatives such as these are present in only very small amounts (there are legal limits on the amounts to be used, see Biocides section) and are essential in providing acceptable product life. Each time the shampoo bottle is opened it receives another dose of micro-organisms.

Preservatives are a feature of all shampoos to prevent bacteria and fungi growth that would result in spoilage of the product, a possible build up of toxic substances and the risk of disease-causing micro-organisms being formed and conveyed to the user. Thus shampoos, in common with many other aqueous formulations, are prevented from biodegradation by the presence of these biocidal preservatives.

There seems to be a dilemma: on the one hand we are looking for more biodegradable chemicals to put in products like shampoos but on the other hand we then go and put preservatives in them to stop bio-degradation. Preservatives are always kept to the minimum – there are legal limits – required to keep the product in good condition for a typical life expectancy and it has to be accepted that at some stage the preservative will fail and the product will degrade.

Colour

Finally there is a colourant identified by its Colour Index (CI) number and present in trace amounts only. CI47005 is quinoline yellow, a bright greenish yellow dye. It is the sodium salt of 2-(2,3-hydroxy-1,3-dioxo-1*H*-indene-2-yl)-6-quinolinesulfonic acid and its 8-quiniline isomer. CI74160 is copper phthalocyanine, a blue pigment (insoluble) which is a complex molecule similar in structure to chlorophyll (see Figure 1.2) but with copper instead of magnesium.

Thus, having studied formulation chemicals we can fully explain the shampoo and even suggest sensible quantities of each chemical ingredient. The formulation, in which the quantities are given on a weight/weight basis, might then look like:

Sodium lauryl ether (3EO) sulfate, 28%	45
Sodium chloride	2
Coconut diethanolamide	2
Coco amido propyl betaine	2
Quaternized protein	1.5
Tetrasodium EDTA	0.1
Citric acid	to pH 5 to 6
Fragrance	qs
Colour	qs
Preservative	qs
Water	to 100

In fact the water content is greater than would appear at first sight because the sodium lauryl ether sulfate also has a water content as does the coco amido propyl betaine. The true total water content is therefore in the region of 80%.

The quantities of each chemical suggested in this simple exercise would need optimizing, by varying the quantities and assessing the outcome, to turn it into an acceptable product.

Whenever formulation work of this nature is carried out it is important to remember that where a particular ingredient has certain properties then incorporating that ingredient into the formulation does not necessarily mean that those properties will be conferred upon the preparation. One ingredient may affect another in a synergistic way and be of benefit to the formulator or the effect may be the exact opposite and be antagonistic. In other words it may be that adding the extra ingredient has a powerful effect out of all expectation, or that it does nothing, or that it is detrimental.

In commercial formulation work the chemist needs to be satisfied that each component is working as best as it can and that there are no unemployed entities – all the chemicals have to earn a living. Synergistic effects often operate in surfactant systems and lead to situations in which the effect of two components together is a greater than the sum of their individual effects. In a complex blend of chemicals there may be many interactions and like this. This is why in formulation work the experimental approach, trial and error, plays a major role.

MACHINE DISH-WASH POWDER

In the above example surfactants were the main functional chemicals. This is the case with many other washing/cleaning preparations and a glance at the labels of selected products such as shower gel, foam bath and manual dish-wash liquids confirms this; these products are alike in terms of ingredients. Car shampoos and wipe over hard surface cleaners also show similarities with these formulations by putting surfactants at the top of the active ingredients list.

However, it is not always the case that cleaning preparations have surfactants in first place. For example a typical machine dish-wash product declares that it is greater than 30% phosphate and less than 5% non-ionic surfactants. Compared with the shampoo surfactants have clearly been relegated. As this is not a product which, in use, involves personal contact there is less onus upon the manufacturer to reveal the contents. And, as there is no personal contact in this cleaning process the conditions and chemicals used may be quite aggressive.

Sodium metasilicate pentahydrate	50
Pentasodium triphosphate	40
Sodium carbonate	5
EO/PO low foam surfactant	3
Sodium dichloroisocyanurate	2

The first thing to comment upon is that the composition is a blend of powders and has no added water. Unlike the shampoo, which was mostly water, this preparation is nearly all active chemical. Of course water is not totally absent as is indicated by the presence of water of crystallization in the sodium metasilicate. Thus the preparation is chemically 'strong' and this, added to the fact that it is used in very hot water, is the basis of an aggressive cleaning system.

Severe conditions are needed because in machine dish-washing the amount of mechanical force and shearing action exerted upon the adhered grease and food residues is small despite it coming from a powerful spray of the wash liquid. However strong the spray, it in no way compares with the forces available in hand dish-washing where, for stubborn soil, an action akin to scrubbing may be used. The machine dish-wash process is almost entirely chemical in its action.

Sodium/Potassium Salts

The greatest proportion of this mixture is a blend of powders which have the combined effect of producing, when the product is dissolved in water, highly alkaline conditions, some buffering action to maintain that level of alkalinity, and sequestering capacity to prevent metal ions such as calcium interfering in the washing chemistry. In addition to the chemical requirements this blend of powders must also provide the correct physical characteristics such as flow and ease of handling. In this respect a major contribution comes from the sodium carbonate.

It is important to recognize here a quantitative aspect, *i.e.* the sodium metasilicate is used as the pentahydrate, $Na_2SiO_3.5H_2O$ with the water of crystallization, amounting to – from simple mole calculation – 42.5%. Thus the amount of active sodium metasilicate is $0.425 \times 50 = 21.3\%$.

EO/PO Surfactant

A consequence of surface activity and the lowering of the surface tension of water is that foam is formed. As was seen when the Gibbs–Marangoni effect was discussed, the stability of a foam depends upon the type of surfactant that is adsorbed at the air/water (gas/liquid) interfaces of the foam cell lamella.

Pure water does not foam because of its high surface tension but, as seen in the shampoo example, a small amount of the right surfactant and a foam is easily generated and has sufficient stability to last. With a machine dish-wash preparation its capacity to form foam must be minimized because of the powerful spray upon the dishes. In this formulation an ethylene oxide (EO)/propylene oxide (PO) block copolymer is used as a low foaming surfactant.

Sodium Dichloroisocyanurate

Although the conditions in machine dish-washing are not conducive to bacterial growth and should in fact destroy most micro-organisms there is always a risk of harmful bacteria remaining. To prevent this becoming a problem a chlorine release agent is used. This compound releases active chlorine to destroy any surviving bacteria, and ensuring that the dishes are, at the end of the process, not only clean but also disinfected.

The above blend of substances is made to attack grease and food residues by chemical action. Such a chemical attack cannot entirely be confined to the dirty dishes but extends to the machine dish-washer itself and the fabric of its construction. To lessen the effect of this corrosion, inhibitors may be added to this type of formulation. One such substance is benzotriazole which is a corrosion inhibitor for the types of alloys found in this type of machine.

CONVENIENCE FOOD

Far removed from shampoo and other washing products we find that much convenience food also contains surfactants. In this type of food, surfactants, natural and man-made, play a secondary but important role and one that is quite different from that in cleaning preparations, although it is still a surface activity role. As with the shampoo we can get some idea of what is happening by scrutinizing the small print on the labels. For the example shown here I had searched through the kitchen cupboards to find a product that is a fair representation of a popular convenience food. A packet soup, vegetable soup, that can be prepared in the cup is the one selected for consideration.

This product is formulated for maximum convenience in that it may sit, dormant, in the cupboard for – if the 'use by date' is any indication – up to eighteen months before being activated by the addition of boiling water. My packet of powdered soup is clearly a very stable product unlike fresh food which is just the opposite. A long life is obviously an essential for this food – as it is for many convenience foods – but it does make one wonder about us choosing to eat old food.

Preventing spoilage of the food during quite long periods of storage is involved and this topic alone can involve a lot of formulation chemistry. For my packet soup, prevention of spoilage by fungi and bacteria is achieved by it being dried and sealed up in an airtight packet thus keeping nature, with its insatiable appetite for biodegradation, at bay.

The product label gives directions for making up the soup and these involve adding 190 ml boiling water to the 80 g of powder from a single packet. As such we get a preparation that is, in effect, about 30% 'active' ingredient and 70% water. And of that 30% much is starch and chemicals. It seems my vegetable soup is a lot of water, a small amount of food and sizeable helping of industrial chemistry.

From the above quantities and other information from the label we get, for the made up soup:

Ingredient	*Chemical name/description*
Water, 70%	
Vegetables, 8%	
Hydrogenated vegetable oil	Vegetable oil reacted with hydrogen to form a fat
Maize starch	Starch, thickening agent
Acidity regulator	An acid salt and buffer, monopotassium phosphate, (potassium dihydrogen orthophosphate)
Emulsifier E471	Monoglycerides and diglycerides of stearic acid
Flavour enhancer E621	Monosodium glutamate
Salt	Common salt, sodium chloride
Colour E104	Quinoline yellow
Flavourings	Sugar, spice extracts

Hydrogenated Vegetable Oil

Vegetable oils, the triglycerides of unsaturated fatty acids, are liquids because of the structure of their fatty acids. Where a fatty acid has a saturated hydrocarbon chain it has a high degree of symmetry that enables it to crystallize easily. However, where the chain contains carbon-carbon double bonds its symmetry is disrupted – in the form of *cis/trans* isomers – and it crystallizes less well. Thus, at room temperature, stearic acid (octadecanoic acid) is a solid whereas oleic acid (*cis*-9-octadecenoic acid) is a liquid. And these properties of the fatty acids are carried over to their glyceride esters.

To get the correct physical properties for the triglyceride the vegetable oil is converted into fat by reacting the double bonds with hydrogen using a supported metal catalyst. Thus, an unsaturated triglyceride, an oil, becomes a saturated fat. As the number of double bonds is less in the hydrogenated product there is less tendency for the product to turn rancid, a reaction in which double bonds become oxidized. As such the hydrogenated product has a longer shelf life. Unfortunately, from a health aspect, this results in turning polyunsaturates into less desirable saturated compounds and converting *trans* fatty acids into their less healthy *cis* isomers.

Starch

In the soup formulation maize starch is used. This is a natural substance and during hydration, when the hot water is added, it forms a thick gel. This not only thickens the preparation but helps stabilize the structure. Other varieties of starch are to be found in convenience foods and are added to produce texture, stability and thickness. Similar in structure and function to starches are the cellulose derivates: methylcellulose E461, hydroxypropylmethylcellulose E464, sodium carboxymethylcellulose E466.

Emulsifier

The basic chemistry in the soup is for the fatty substance, the hydrogenated vegetable oil, to be made into an o/w emulsion upon adding hot water and stirring. The importance of an emulsion lies in the fact that some of the flavour components are fat soluble and some are water soluble.

Without emulsification the soup would simply exist as an aqueous layer upon which floats an oily layer from the melted fat. Apart from looking unappetizing each layer would have a different texture and taste. What is required is a dispersion so that all the flavour components are spread throughout the bulk and there is a consistent texture.

The dispersion is an o/w emulsion formed by the emulsifying agent, a non-ionic surfactant consisting of the ester of glycerol and fatty acids. As seen in the section on emulsification a mixture of surfactants providing the correct HLB value for the oil is needed, *i.e.*, surfactant molecules with just the right proportion of hydrophilic and lipophilic properties. Glyceryl stearates, in which the hydrophile is the remaining hydroxyls in the glycerol molecule and the lipophile is the hydrocarbon chain of the stearic acid, fulfils the role and is also assisted by the thickening effect of the starch which helps prevent the oily particles coalescing, which would lead to the emulsion breaking and complete phase separation.

Surfactants figure in many convenience foods, usually as emulsifiers and form the E400 series in the E Number listings. Some other examples of food surfactants are the mono and diglyceride esters of acetic, lactic, citric and tartaric acids. Recently sucrose esters (E473) have come onto the market as food additives. The hydrophile in these surfactants is the sucrose molecule and the lipophile is a fatty acids as shown in Figure 3.1. The HLB value of these can be adjusted by either changing the number of fatty acids attached to the hydroxyl oxygens or altering the length of the fatty acid hydrocarbon chains. As such these sucrose esters are very flexible in terms of their properties.

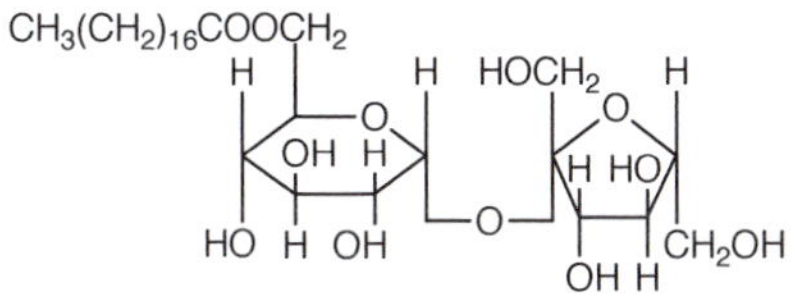

Figure 3.1   *Sucrose esters, used as surfactants in some foods, in which both hydrophile and lipophile are from plant sources. Sucrose monostearate is shown here. The other hydroxyl groups can also be reacted with fatty acids to produce di-esters, tri-esters* etc. *leading to a wide range of HLB values*

In general, the permitted surfactants are simple chemical modifications of natural, mainly plant, materials. It is surprising therefore to come across a series of ethoxylated compounds such as the polysorbates. Polysorbate 20, E432, is polyoxyethylene (20) sorbitan monolaurate. It has a hydrophile composed of a chain of twenty ethylene oxide (EO) groups which makes most of the molecule synthetic and derived from petroleum.

One compound that is entirely natural is lecithin, E322, which is a mixture of diglycerides of stearic, palmitic and oleic acids joined to a phosphatide as shown in Figure 3.2. Lecithin is commonly used in margarine and chocolate where it exhibits both surface active properties and works as an anti-oxidant.

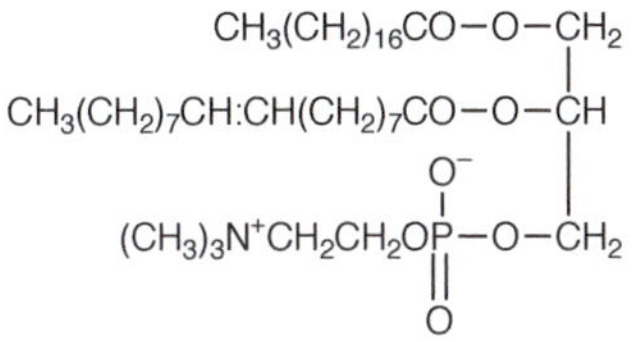

Figure 3.2   *Lecithin is entirely natural and comprises a mixture of glycerides of stearic, palmitic and oleic acids joined to a phosphatide group*

Surfactants other than emulsifying agents are also to be found in prepared foods. For example, sodium dioctyl sulfosuccinate, a synthetic
anionic wetting agent, is permitted in certain foods; it is not that the
particular food needs a wetting agent but that the surfactant is present in
one of the ingredients where it is used as a processing aid.

Acidity Regulator

It is important that the pH does not change as this may result in changes
in flavour and texture. Colourants can be particularly sensitive to pH
excursions giving rise to a completely different colour; in effect they
behave as acid/base indicators. To ensure the pH does not drift, potassium dihydrogen orthophosphate is incorporated. This is an acid salt and
buffer and as such guards against pH changes.

Salt

Salt, sodium chloride, is also present but unlike the shampoo where it
was there to thicken the surfactant here its main function is for taste.

Colours

Quinoline Yellow, a synthetic dye, is used here just as it was in the shampoo studied above. Colourants, dyestuffs and pigments, are essential
ingredients in many convenience foods. Often the original food, particularly vegetables, used in making these preparations soon loses its colour
and would, without added colourants, appear unappetizing. Furthermore large amount of other ingredients such as those reported above
have little or no colour.

Thus, to obtain a product that has a colour that fits in with the consumers perception of the product colourants are needed. Sometimes
these can be quite vivid, far exceeding the colour intensity of the original
food and have a spectrum that has a decided lean towards the ultraviolet. For example, the contents from the can labelled mushy peas bor-
der on being chemiluminescent. This interesting visual effect is achieved
using the colourants Tartrazine E102 and Brilliant Blue E133. Another
blue colourant used in foods, an insoluble blue pigment, is Copper
Chlorophyll E141; this is similar to the Copper Phthalocyanine CI74160
used in the shampoo.

Substances used in foods and food packaging may well be chemically
identical to those found in other areas of formulation work but for food
use composition and purity must comply with strict specifications.

PHARMACEUTICALS

Pharmaceutical preparations cover a wide range with many different bio-active substances and a host of other chemicals that are essential for effective application. Among these chemicals surfactants appear frequently for carrying out the following functions: wetting, solubilizing and emulsifying. Thus, a glance at the pharmaceutical literature gives: the non-ionic surfactant Polysorbate 20, (polyoxyethylene 20 sorbitan monolaurate) as an oil-in-water emulsifier in Liquid Peppermint Emulsion; Polysorbate 80, (polyoxyethylene 20 sorbitan mono-oleate) in Coal Tar Ointment; the sodium oleate soap formed, *in situ*, by reaction between oleic acid and ammonia, as an emulsifying agent in White Liniment; the anionic surfactant, sodium lauryl sulfate, combined with cetostearyl alcohol as an emulsifying wax base; chloramphenicol, the antibiotic, is solubilized by the micelles formed from ethoxylated sorbitan esters.

Sodium lauryl sulfate appears in other pharmaceutical products where it is useful at promoting dissolving. Diltiazem hydrochloride, an anti-anginal drug comes in capsule form in which sodium lauryl sulfate is one of the ingredients. In each of these examples the surfactant plays a supportive role rather than being the active ingredient.

However, some preparations do have surfactants as active components, for example, the cationic surfactant cetylpyridinium chloride is used for its bactericidal action in mouthwash for dental hygiene. Benzalkonium chloride, another cationic bactericide, also figures in eye bath liquid for rinsing the eyes to alleviate one of our more recent afflictions, tired eyes – presumably the result of too much looking.

There is currently much research into the different ways in which drugs can be delivered to the precise location where their bio-activity is required. This site-specific targeting relies upon the particular drug molecules being transported by some kind of vehicle to the site from some distant point of entry into the body. An important aspect of this transport is that the bio-active molecules arrive intact, *i.e.*, they are not damaged or wastefully consumed *en route*. Here, surfactants can also play a role in that, due to their ability to form micelles and solubilize lipophilic molecules within the micelle, they can act as vehicles.

Thus a drug molecule may be safely locked into a micelle until the desired destination is reached at which point, due to some clever chemistry, the micelle breaks open and releases its cargo, the drug molecule, in perfect condition and ready for action.

The surface active properties of the ethylene oxide/propylene oxide block polymers have an important role here. They can carry, say, a lipophilic anti-cancer agent, held inside a micelle by means of the intermolecular attractions between the surfactant lipophile and the drug molecule, to a cancerous tumour whilst the outer hydrophilic layer protects it.

Gemini surfactants are capable of producing very low surface tensions in solution that are of interest in drug delivery research. Furthermore these surfactants are undergoing research as devices for delivering genes through the cell membrane and all the way into the nucleus.

MISCELLANEOUS

There are other formulations which are a part of modern day living that rely upon surfactants as active ingredients or as substances playing a supporting role to improve the effectiveness of others. In surface coating preparations it is logical that surface chemistry is going to play a big part. To make a preparation, paint or ink, to be applied to a substrate requires a study of wetting, spreading, viscosity and foam formation. And this means an understanding of surfactants.

As many of the solvent-based coatings with their high levels of Volatile Organic Compounds (VOCs) are giving way to waterborne coatings surfactant chemistry becomes even more important. Over recent years special aqueous-based inks have been developed for the ink-jet printer. This is one area in which the high-tech specifications impose a real demand upon ink formulation and to meet that demand surfactant technology is playing a major part.

The above examples serve to demonstrate the importance of formulated products that we come across on a daily basis. However, the story does not finish here for there are even more formulations – and a lot of them use surfactants – to be found in various industries: agricultural pesticides, minerals refining, oil drilling, road surfacing, soil decontamination, fuels, lubricants, washing/degreasing, textiles and dyeing.

Chapter 4

Further Formulations

In this chapter we will take a general look at a wide range of formulations for familiar preparations. The formulations are shown for guidance only; they are not a recommendation to manufacture but they may be suitable as starting points for anyone wishing to develop a particular product. Where the latter is the case then advice should be sought from the chemical suppliers before starting and it should be borne in mind that if a preparation is made for sale to the public there are many other considerations and responsibilities.

Each of the formulations has been selected as being suitable for small scale trial preparations in the laboratory using normally available equipment, namely: balance, variable speed stirrer, steam bath, thermometers, pH meter and buffer solutions, measuring cylinders, beakers, personal protection equipment *etc.* A few of the emulsion preparations may require a high shear mixer such as a Silverson.

For personal care products ingredients specifically rated for such use must be used along with deionized water and high standards of cleanliness of the mixing equipment. Furthermore, there must be thorough testing such as effectiveness, stability and bacteriological tests, to ensure the product is fit for its intended purpose.

PERSONAL CARE PRODUCTS

Products formulated for personal care applications are clearly going to have to strict limitations on their properties. A mixture of chemicals for putting onto the hair and skin is occasionally going to find its way into the eyes and small amounts of it may even be ingested. These possibilities must be taken into consideration but tempered with a degree of common sense. Using regular formulation practices along with personal care grade ingredients from a reputable supplier is the first step in ensuring safe preparations.

For personal care products important requirements, apart from the need for the product to perform its intended task, are that it must work at 40 °C and less, that it must have a pH that is neutral to slightly acidic and that irritancy to eyes and skin is minimized. Irritancy is usually caused by the unassociated surfactant molecules rather than surfactant in the micelles. Thus, high concentrations of surfactant monomer, from highly soluble surfactants, producing high CMCs, can be quite irritant.

Careful choice of surfactant type and combinations of surfactants can minimize this. In some cases it is possible to combine surfactants so that the more soluble one, the more irritant one, is held in a mixed micelle. Mildness is desirable in all personal care products but in baby products, such as shampoos, it is essential. The growth of amphoteric surfactants in personal care products is due in the main to meet the demands for mildness.

Cationics, quaternary ammonium compounds, are a popular ingredient in hair care and skin care products because of their capacity to form an ionic bond with proteins and impart desirable properties such as providing a soft feel to skin or making hair glossy and anti-static. Cationic surfactants react with anionic surfactants to produce an inactive fatty molecule and so care needs to be exercised when cationics are to be combined with other surfactants.

To prevent excessive degreasing of the skin super-fatting additives are often put into shampoos, shower gels and bath products. Lanolin and its derivatives have proven to be particularly popular here. Pearlizing agents put into these products are insoluble substances and will eventually separate and either float or sink unless the product is of relatively high viscosity.

As most personal care products are used a bit at a time and may spend lengthy periods on the bathroom shelf they could become media for the growth of bacteria and fungi. And each time bottle or jar is opened even more micro-organisms are introduced from the air and hands. An essential requirement of personal care products is the appropriate biocidal preservative.

1 Conditioning Shampoo

Here the anionic surfactant is triethanolamine lauryl sulfate which has good solubility characteristics that provide for easy rinse off. The hydrolysed vegetable protein derivative is substantive to the hair protein and so improves condition. Lactic acid ensures slight acidity to prevent irritancy to the skin.

Water	to 100
Triethanolamine lauryl sulfate 40%	35.00
Coconut diethanolamide	2.00
Coco amido propyl betaine 30%	10.00
Hydrolysed vegetable protein	1.00
Sodium chloride	qs
Lactic acid	to pH 6.0
Perfume, colour, preservative	qs

The sodium chloride is dissolved in a small amount of the water. The remainder of the water is warmed to 60 °C and other ingredients, excluding the perfume and lactic acid, are stirred in until fully dissolved and then the sodium chloride solution is added and stirring continued as viscosity rises. Lactic acid is added whilst the blend is still warm and addition continued until the pH stays constant at 6.5 at near room temperature; the perfume is also added at this point.

2 Shampoo for Dry Hair

In this shampoo the total surfactant actives is $(0.4 \times 8) + (0.27 \times 6) + (0.7 \times 4) + (0.3 \times 7) + 2 = 11.7\%$.

Triethanolamine lauryl sulfate 40%	8.00
Ammonium lauryl sulfate 27%	6.00
Sodium lauryl ether sulfate 70%	4.00
Coco amido propyl betaine	7.00
Coconut diethanolamide	2.00
Citric acid	to pH 5.5–6.0
Sodium chloride	qs
Perfume, colour, preservative	qs
Water	to 100

The water is warmed to about 50 °C and the sodium lauryl ether sulfate dissolved into and, when homogenous, followed by the two lauryl sulfates and the diethanolamide. When below 35 °C the perfume, colour and preservative are added and the pH adjusted with the citric acid. Finally, the viscosity is adjusted by means of sodium chloride.

3   Shampoo for Greasy Hair

To cope with the extra grease from the hair these types of shampoo contain a slightly higher levels of surfactant and/or more powerful degreasing ones than are used in regular shampoos. Here the total surfactant actives is $(0.4 \times 5) + (0.27 \times 7) + (0.7 \times 8) + (0.3 \times 7) + 1 = 12.6\%$.

Triethanolamine lauryl sulfate 40%	5.00
Ammonium lauryl sulfate 27%	7.00
Sodium lauryl ether sulfate 70%	8.00
Coco amido propyl betaine	7.00
Coconut diethanolamide	1.00
Citric acid	to pH 5.5–6.0
Sodium chloride	qs
Perfume, colour, preservative	qs
Water	to 100

The procedure is similar to that for the previous formulation.

4   Pearlized Shampoo

The pearl effect in shampoos and bath/shower preparations is brought about through the addition of ethylene glycol monostearate which stays as an insoluble dispersion. Crystalline platelets of this fatty ester reflect and split the light to give the silvery rainbow hue that provides the pearl appearance.

Water	64.5
Lanolin ethoxylate	5.0
Sodium lauryl ether sulfate 50%	20.0
Alkyl amidopropylbetaine 30%	7.0
Coconut diethanolamide	1.5
Ethylene glycol monostearate	1.0
Sodium chloride	1.0
Perfume, colour, preservative	qs

The sodium chloride is dissolved in a small amount of the water. The remaining volume of water is heated to 70 °C and other components, except for the perfume, are then blended in and the sodium chloride solution added. When cool the perfume is stirred in.

5 Conditioning Rinse

This is used after washing the hair with a regular shampoo that does not contain its own conditioner. It has several functions: the acid causes the surface of the hair shaft to return to its natural state whilst the hydrophiles of the cationic surfactant bond to the hair protein. This results in a hydrophobic layer the properties of which are improved by the fatty alcohol.

Water	91.0
Cetrimonium methosulfate	3.0
Coconut monoethanolamide	2.0
Cetyl/stearyl alcohol	4.0
Citric acid	to pH 2.5–3.0
Perfume, dye, preservative	qs

The water is brought to 70 °C and the waxes stirred in to form a smooth blend that is then allowed to cool before addition of the other ingredients and adjustment of pH.

6 Foam Bath

An important feature of foam bath preparations is an ability to exhibit flash foam characteristics and good foam stability. In the preparation shown below a blend of anionic and amphoteric surfactants is used to achieve this. The amphoteric excels in its ability to generate foam.

Sodium lauryl ether sulfate 27%	60.0
Lauryl betaine 30%	5.0
Sodium chloride	qs
Citric acid	to pH 6.5 to 7.0
Perfume, colour, preservative	qs
Water	to 100%

A small volume of the water is kept separate for dissolving the citric acid. Into the bulk of the water the sodium chloride is dissolved followed by the other ingredients except for the citric acid. Dissolved in the small volume of water the citric acid is then added until the target pH is reached.

7 Soluble Bath Fragrance

Solubilization of an essential oil by a non-ionic surfactant to form a clear micro-emulsion is demonstrated by this formulation. Note the relatively large amount of surfactant compared with the oil. This is needed to ensure that the oil becomes trapped within the hydrophobic interior of the non-ionic micelles to produce a clear colloidal dispersion.

With a lesser proportion of surfactant a regular emulsion would be formed and the product be cloudy due to larger particles sizes that then scatter the light. In use the micelle structure breaks down as it hydrates in the large volume of water; this releases the oil which provides a thin, slowly evaporating, film on the surface of the bath water to produce a constant fragrance.

Water	70.0
Sorbitan monolaurate (20EO)	25.0
Fragrance oil	5.0
Colour, preservative	qs

A blend of the surfactant and oil is made and then added, with stirring, to the water followed by colour and preservative.

8   Shower Gel

Again, sodium lauryl ether sulfate and coconut diethanolamide feature as the important actives in many shower gel formulations such as the one shown here. Sodium chloride is used to increase thickness by causing the surfactant to restructure into the high viscosity cylindrical micelle structures. Lactic acid, a natural acid and thought to be a component of the skin, is used to make the pH compatible with that of the skin.

Water	to 100
Sodium lauryl ether sulfate 30%	40.0
Coconut diethanolamide	2.0
Alkyl amido propyl betaine 30%	5.0
Cocoamine oxide	2.0
Sodium chloride	1.0
Perfume, colour, preservative	qs
Lactic acid	to pH 6.5

The sodium chloride is dissolved in a small amount of the water. The remainder of the water is warmed to 60 °C and other ingredients, excluding the perfume and lactic acid, are stirred in until fully dissolved and then the sodium chloride solution is added and stirring continued as viscosity rises. The lactic acid is added whilst the blend is still warm and addition continued until the pH stays constant at 6.5 at near room temperature; the perfume is also added at this point.

9   Roll-on Antiperspirant

Here the surfactant, ethoxylated lanolin, has several roles to play: emollient, solubilizer for the perfume, humectant to absorb and retain moisture thus preventing drying out of the aluminium compound that would form crystals and block the ball. Ethanol enables rapid evaporation once the antiperspirant has been delivered to the skin. Some stabilizing thickness is obtained from the cellulose derivative. Antiperspirant activity comes from the aluminium chlorhydrate.

Water	37.35
Hydroxyethylcellulose	0.65
Ethanol	20.00
Polyethoxylated lanolin	2.00
Aluminium chlorhydrate 50%	40.00
Perfume, colour, preservative	qs

The hydroxyethylcellulose is dispersed in the alcohol; it does not dissolve at this stage. The water is heated to 60 °C and the lanolin derivative blended in followed by cooling to room temperature. To the cool liquid is added the pre-dispersed thickener with constant stirring but avoiding entrapping air bubbles. As the preparation gels the aluminium chlorhydrate and perfume are slowly added.

10   Baby Cleansing Lotion with Lanolin

Combining two surfactants with different HLBs gives the desired HLB (9 to 10) for good emulsion forming properties in this oil-in-water emulsion formulation. Hydroxyethylcellulose stabilizes the emulsion due to thickening of the aqueous phase. During application the emulsion breaks and leaves a thin film of the waxy substances to act as water repellent. The propylene glycol acts as humectant and retains a small amount of moisture to prevent the skin being too dry.

Water	69.4
Heavy liquid paraffin	20.0
Cetyl/stearyl alcohol	1.0
Pure liquid lanolin	1.0
Sorbitan monostearate (HLB 4.7)	2.0

Sorbitan monostearate (20EO) (HLB 14.9)	2.0
Hydroxyethylcellulose	0.5
Propylene glycol	4.0
Perfume, preservative	qs

The hydroxyethylcellulose powder is vortexed into the cold water and given time to thicken after which the aqueous phase and oil phase, which have been separately heated to 65 °C, are blended by adding the hot oil to the aqueous phase whilst stirring. After cooling to less than 40 °C the perfume is added.

11   Baby Cream

Protecting the baby from nappy rash and chafing is the primary function of this preparation. This water-in-oil emulsion, after application, breaks as the water evaporates producing a protective waterproof hydrocarbon barrier on the skin. The main hydrocarbon in this formulation is petroleum jelly rather than mineral oil as is used in the thinner preparations such as lotions and milks. Sorbitan isostearate is used as the surfactant due to low toxicity, skin mildness and, being based on a saturated fatty acid lipophile, is not prone to oxidation. A dispersion of zinc oxide assists in reducing inflammation of the skin.

Water	57.40
Sorbitan isostearate	2.00
White petroleum jelly	20.00
Microcrystalline wax 3520	3.50
Liquid Paraffin 25cS	7.35
Glycerol	2.00
Magnesium sulfate heptahydrate	0.70
Zinc oxide	7.00
Perfume, preservative	qs

The oils and waxes are heated to 70 °C and blended together. Water, glycerine and magnesium sulfate are mixed and heated to 70 °C and the hot oil added to this solution whilst stirring. Zinc oxide is blended in at 60 °C and when below 40 °C the perfume is added. The final cold preparation may require further treatment by passing through a homogenizer.

12   Liquid Soap

Most liquid soaps are in fact detergents and not soap. Usually filled into a pump-action dispenser this makes these products more hygienic and less messy than bar soap. Liquid soaps also have the advantage that ingredients can be added that would not be possible in bar soap. As such these products offer plenty of formulation flexibility. However, the extra packaging of the throw away dispenser does detract from some of those advantages on environmental grounds.

Sodium lauryl ether sulfate 27%	20.0
Monoethanolamine lauryl sulfate	10.0
Lauryl betaine 30%	7.0
Sodium chloride	1.0
Citric acid	to pH 6.5–7.0
Perfume, colour, preservative	qs
Water	to 100%

The sodium chloride is dissolved in the water and the other ingredients, with the exception of the citric acid, blended in until a smooth homogenous mix results. Finally the citric acid, dissolved in a small amount of water, is added until the required pH is obtained.

13   Toothpaste

Here a polishing agent, dicalcium phosphate, is dispersed in a solution of glycerol, water and surfactant. The latter functions as a wetting and emulsifying agent to remove grease and food residues from the teeth. Sodium carboxymethylcellulose (SCMC) is the thickener and binder and prevents the polishing agent from settling out. Gum tragacanth, a natural chemical, may be used in place of the SCMC. Sodium monofluorophosphate is added (there is a statutory maximum) to provide the low concentration of fluoride that is thought to be beneficial in improving tooth enamel by converting the apatite into fluoroapatite.

Water	21.50
Glycerol	25.00
Sodium carboxymethylcellulose	1.00
Sodium monofluorophosphate	0.80
Sodium saccharin	0.20
Dicalcium phosphate dihydrate	50.00
Sodium lauryl sulfate	1.50
Flavour, preservative	qs

The gum is mixed with the water and allowed to thicken. The other ingredients, apart from the dicalcium phosphate, are then mixed into the water. Finally the dicalcium phosphate is blended in to form a smooth paste.

14   Barrier Cream

In many industrial processes the hands come into contact with sub-stances that can irritate or damage the skin. To prevent this a strongly water repellent type of hand cream, known as barrier cream, is formu-lated. Once applied and well worked in the water starts to evaporate and the emulsion breaks. As a consequence a continuous wax film is spread over the surface of the skin as a hydrophobic coating. Careful choice of the waxy blend gives the film good adhesion properties and the required flexibility.

Water	57.0
Glyceryl monostearate self-emulsifying	11.0
Cetyl/stearyl alcohol	2.2
Sodium lauryl sulfate	0.8
Beeswax	4.0
Lanolin	6.0
Glycerol	4.0
Zinc stearate	15.0
Perfume, dye, preservative	qs

The sodium lauryl sulfate and glycerol are dissolved in the water and heated to 70 °C. The glycerol monostearate, cetearyl alcohol, beeswax, lanolin and zinc stearate are mixed and slowly brought to 70 °C. When completely homogenous the wax phase is blended with the aqueous phase and, with constant stirring, allowed to cool to 35 °C at which point the perfume, dye and preservative are added.

15   Cleansing Beauty Milk

Another oil-in-water emulsion is this milk in which the emulsifying agent is a soap formed *in situ* by the neutralization reaction between stearic acid and triethanolamine to give triethanolamine stearate.

Water	16.20
Lanolin alcohol	1.10
Cetyl alcohol	0.40
Heavy liquid paraffin	8.40
Stearic acid	4.20

Triethanolamine	2.10
Perfume, colour, preservative	qs

Triethanolamine is dissolved in the water and brought to 70 °C. The other ingredients with the exception of the perfume are also heated to this temperature followed by addition of the water phase to the oil with stirring. Perfume is added when cool.

16   Natural Moisture Cream

This water-in-oil non-ionic emulsion uses two emulsifiers, polysorbate 60 and sorbitan stearate, for maximum stability. The other esters give a soft emollience to the skin and improve application properties. Hydrolysed vegetable protein produces conditioning and moisturising properties.

Oil phase	
Caprylic/capric triglyceride	13.0
Octyl cocoate	3.0
Cetyl esters	3.0
Cetyl/stearyl alcohol	3.0
Polysorbate 60	3.0
Sorbitan stearate	2.0
Aqueous phase	
Water	69.0
Glycerol	3.0
Hydrolysed vegetable protein	1.0
Perfume, colour, preservative	qs

The oil and water phases (except for the perfume and hydrolysed vegetable protein) are kept separate and heated to 65–70 °C. The oil phase is then added to the aqueous phase with stirring until the temperature is below 40 °C at which point the perfume and hydrolysed vegetable protein are added. The oil phase is then added to the water phase. The preparation is filled off at 30 °C.

17   Insect Repellent Cream

Active components may be added to a simple emulsion to provide certain functionality as is demonstrated in this preparation in which menthol, diethyl toluamide and allantoin are incorporated into an emulsion.

Water	to 100
Higher fatty alcohol ethoxylate wax	2.00
Glyceryl monostearate	3.00
Diethyl toluamide	25.00
Menthol	0.25
Allantoin	0.25
Carbopol carboxy polymer thickener	0.25
Ethanol	2.00
Triethanolamine	to pH 6.5
Perfume, preservative	qs

The thickener is dispersed in the water with stirring until thick followed by addition of the ethanol. Both the aqueous and oil phases are separately heated to 65 °C and the aqueous phase added to the oil with stirring. Stirring is continued whilst cooling during which time the perfume is added and the pH adjusted to 6.5 when the preparation is about 20 °C.

18   Aerosol Shaving Foam

Emulsification of the paraffin is brought about by triethanolamine stearate/myristate soaps that are formed *in situ*. Lanolin assists the emulsification and acts as a super-fatting agent preventing dry skin and the propylene glycol functions as a humectant to retain moisture. Being an aerosol some environmental points are forfeited but this may be regarded as acceptable for a product that is ready to use rather than time spent in preparing a foam by more traditional methods.

Water	83.50
Lanolin liquid	2.00
Myristic acid	2.00
Stearic acid	5.00
Liquid paraffin 25cS	1.00
Propylene glycol	3.00
Triethanolamine	3.50
Perfume, preservative	qs
Aerosol: base material plus butane propellant	

The triethanolamine and propylene glycol are dissolved in the water which is then heated to 70 °C. The other components with the exception of the perfume are also heated to this temperature after which the aqueous phase is added to the oil phase with stirring which is continued during cooling. At below 40 °C the perfume is added.

HOUSEHOLD PRODUCTS

In general, for the cleaning products discussed here there are not the constraints that apply for personal care preparations but it must be borne in mind that for some of the products there is going to be skin contact. Many of the cleaning formulations outlined below work in conditions where hot water, high levels of alkalinity and strong agitation are not a problem.

The stability of chemicals in hot alkaline conditions must be taken into consideration. For example, some thought needs to be paid to cloud point when using non-ionic ethoxylates in formulations destined for use in hot water. Most surfactants are tolerant of high alkalinity but the same cannot be said for acid conditions. Acid hydrolysis breaks down alkyl sulfates, sulfosuccinates and alkyl polyglucosides. Some formulations in this section are concentrates to be diluted by the user. Where this is the case water hardness needs some serious consideration. If the user is in a hard water area there must be sufficient sequestrant to allow for this.

In conditions of high alkalinity or high acidity the problems of bacterial or fungal growth are less but for preparations having less severe conditions, as with the personal care products, preservatives may be a requirement.

Dish-wash Liquids

In hand dish-washing liquids there can be lengthy periods of skin contact and so mildness is essential. Viscosity needs to be such as to enable about the right amount of detergent to be dispensed from a squeeze bottle and, once mixed into the wash-up water, good foam with sufficient stability to last throughout. Typically there are three qualities of dish-wash liquid, each reflecting the amount of active matter, from the cheap ones to the premium products; the latter representing good quality, best value and least damaging to the environment.

19   Dish-wash Liquid 10% Active

Water	to 100
Sodium dodecylbenzene sulfonate 60%	11.70
Coconut diethanolamide	1.00
Sodium lauryl ether sulfate 27%	6.60
Sodium chloride	qs
Perfume, colour, preservative	qs

A small amount of the water is taken to dissolve the sodium chloride. To the rest of the water are added the other ingredients and stirring carried out until a homogenous blend results. Finally the sodium chloride is added to give the desired viscosity.

20   Dish-wash Liquid 20% Active

Water	to 100
Sodium dodecylbenzene sulfonate 60%	23.3
Coconut diethanolamide	2.00
Sodium lauryl ether sulfate 27%	13.30
Sodium chloride	qs
Perfume, colour, preservative	qs

The mixing details are as for the previous formulation.

21   Dish-wash Liquid 30% Active

In this formulation the anionic surfactant, sodium dodecylbenzene sulfonate is produced *in situ* by neutralization of the sulfonic acid with sodium hydroxide. This neutralization, and others alike, require a good degree of precision to ensure that excess alkali or acid, which would lead to extremes of pH, does not arise.

Being a fairly concentrated solution it is necessary to use ammonium lauryl ether sulfate in place of the less soluble sodium salt. In addition a hydrotrope is required to keep down the viscosity to a reasonable level and this functions by taking part in the micelle structure and limiting the extent of sphere-to-rod transitions. Of course, no sodium chloride is required in this one.

Water	to 100
Sodium hydroxide 47%	4.03
Dodecyl benzene sulfonic acid	14.10
Sodium xylene sulfonate	6.00
Coconut diethanolamide	2.00
Ammonium lauryl ether sulfate 60%	24.26
Perfume, colour, preservative	qs

The mixing involves neutralization of the sulfonic acid by sodium hydroxide and, as is common with this type of reaction, there is an appreciable exotherm. As such this mixing process demands rather more attention to procedure than many other formulations. The sodium hydroxide is added to the whole of the water with stirring. When fully dispersed the sulfonic acid is added slowly until homogenous.

The other ingredients, with the exception of the ammonium lauryl ether sulfate, are added and the pH adjusted to 7.0 by small supplementary additions of the sulfonic acid or diluted sodium hydroxide. If the ammonium compound were added before pH adjustment there is the risk of it going into an alkaline mix which would result in it giving off ammonia and being converted into the less soluble sodium salt.

22 Air Freshener Gel

This preparation should appeal to those looking for formulations that use natural chemicals. Carrageenan, the thickener and gel stabilizer, is extracted from seaweed, the surfactant is made (part synthetic) from the seeds of the castor oil plant and the fragrance oil gives the option of natural oils.

The gel is a clear micro-emulsion in which the fragrance oil is solubilized by being incorporated into the hydrophobic interior of the large micelle structures formed from the surfactant. As the water slowly evaporates the micelle structure at the gel/air interface becomes unstable, bursts and releases the fragrance molecules. Thus, a continuous slow release of fragrance is achieved and very little solid residue remains.

Water	87.00
Carrageenan	3.00
Fragrance oil	5.00
Hydrogenated castor oil (50EO)	5.00
Preservative, colour	qs

During the blending it is important to avoid air being trapped in the gel. The colour and preservative are added to the water and the carrageenan dispersed and left stirring whilst slowly heating to 70 °C for the thickness to develop. Once thick and homogenous the aqueous phase is allowed to cool. Fragrance oil and surfactant are thoroughly blended and added to the aqueous phase when it has dropped to 55–60 °C. When homogenous the blend may be filled off into containers before further cooling produces the immobile gel.

23 Heavy Duty Degreaser

The major ingredient here is the natural product d-limonene extracted from the peel of citrus fruits which is both environment friendly and from a renewable resource. Oleic acid from natural substances such as olive oil and the citric acid also score environmental points. Emulsification of the limonene is primarily by the soap formed due to neutralization of the oleic acid by the sodium hydroxide. The glycol ether is a water-soluble solvent that is generally regarded as one of the healthier man-made solvents.

Water	to 100
d-Limonene	54.3
Oleic acid	5.00
Tridecyl alcohol ethoxylate (7EO), 85%	15.00
Dipropylene glycol mono methyl ether	10.00
Sodium hydroxide 50%	0.30
Citric acid	to pH 7

The oleic acid, alcohol ethoxylate, glycol ether and limonene are blended and warmed to 50 °C. The sodium hydroxide is added to the water and the resulting solution warmed to 50 °C and then added to the oil phase with stirring. When homogeneous and near to room temperature the citric acid is added, given time to dissolve, and the pH tested. This is continued until a pH of 7 is reached.

24   Furniture Polish, Plant Wax

Furniture polish is traditionally based upon plant waxes such as car-nauba wax or animal waxes such as beeswax and is often prepared as emulsions. Emulsification is usually by means of the soap formed by neutralization of a oleic acid, a fatty acid, with a base such as morpho-line. Following application to the surface, the water evaporates, the emulsion breaks and the wax/soap particles fuse to form a hard film that can be buffed to a brilliant gloss.

Water	82.0
Carnauba wax	13.7
Oleic acid	2.7
Morpholine	1.6

The wax and oleic acid are heated to 90 °C to produce a homogenous melt into which the morpholine is added. The water is heated to boiling and combined with the wax melt. Rapid cooling is carried out whilst stirring constantly.

25 Shoe Cream

A combination of paraffin wax and fatty acid waxes are maintained as an emulsion in this preparation which produces gloss by buffing of the partially dried cream to give the gloss. Dyes and pigments could be added as desired to tailor the product to a particular leather colour. Shoe polishes, apart from cleaning and adding gloss, also protect the leather by lubricating the fibres in the leather matrix.

Fatty acid triglyceride C18–36 synthetic wax	6.0
Fatty acid C18–36 synthetic wax	2.6
Paraffin wax 140/145	8.0
Cetyl/stearyl alcohol	1.2
White spirit	28.0
2-aminomethylpropan-1-ol	1.2
Water	53.0
Preservative	qs

The waxes and the cetearyl alcohol are blended together at 90 °C, allowed to cool to about 80 °C and the white spirit mixed in. The water, 2-aminomethylpropan-1-ol and preservative are heated to 80 °C and the phase added with high shear stirring. Stirring is continued whilst the blend cools and thickens down to about 50 °C.

26 Furniture Polish, Siliconized

It is common practice to use a combination of waxes to optimize gloss and other factors. The example shown here uses two natural waxes and the synthetic silicone along with a non-ionic surfactant as emulsifier and a slow evaporating solvent. Once applied, the water evaporates to leave a plastic film of the waxy solids dispersed in the solvent. This is readily buffed to a shine during which process the solvent evaporates leaving a hard gloss.

Water	81.0
Carnauba wax	2.0
Beeswax	2.0
Polydimethylsiloxane silicone 350cS	2.0
Sorbitan tristearate blend	3.0
White spirit	10.0

The waxes, silicone, white spirit and surfactants are heated to about 90 °C. Water just below boiling point is slowly added to the wax melt with constant stirring. When homogenous the blend is rapidly cooled whilst stirring is maintained.

27   Metal Polish

Reaction between the triethanolamine, ammonia and oleic acid produces the emulsifier which, with the white spirit, gives a thick emulsion to disperse the abrasive.

Water	to 100
Oleic acid	2.5
Ammonia solution 26 Be	1.9
Triethanolamine	0.6
White spirit	36.0
Abrasive	16.0

The oleic acid, triethanolamine, white spirit and abrasive are blended together and, with slow stirring, the water and ammonia are added.

28   Machine Dish-wash Powder

This one was considered in the previous chapter but is reported again to outline the blending procedure.

Sodium metasilicate pentahydrate	50
Pentasodium triphosphate	40
Sodium carbonate	5
EO/PO low foam surfactant	3
Sodium dichloroisocyanurate	2

This is essentially a powder blend formulation and requires some attention to detail in the mixing process to ensure there are no pockets of localized concentration. The blending of powders results in a heterogeneous mix, unlike the blending of liquids where a homogenous product is normally formed. With mixed powders there is also the risk of separation during handling and this calls for careful attention when considering particle size and bulk density of the ingredient powder.

29   Oven Cleaner

A formulation with strongly caustic properties required to attack burnt on grease is required for oven cleaning. In this preparation caustic potash, potassium hydroxide, is used. Hydroxyethylcellulose acts as thickener to enable the solution to be retained on vertical surfaces. Just about all of the grease is from animal and vegetable origins and consists of triglycerides which are saponified by strong alkalis, hence the relatively high concentration of caustic.

Water	77.50
Fatty alcohol ethoxylate low foam	3.00
Trisodium nitrilotriacetate, 38%	2.50
Sodium cumene sulfonate	8.00
Tetrapotassium diphosphate	2.00
Potassium hydroxide	4.00
Hydroxyethylcellulose	3.00

The potassium hydroxide is first dissolved in the water followed by the tetrapotassium diphosphate. Once dissolved the other components, except for the hydroxyethylcellulose, are blended in. Finally the hydroxyethylcellulose is slowly added by sprinkling and the whole stirred until thick and homogenous.

30 Cream Hard Surface Cleaner

The anionic surfactant sodium dodecyl sulfonate is an efficient hard surface cleaner. Although available as such it is often convenient to produce it *in situ* by neutralizing the acid with sodium hydroxide and this is demonstrated in the formulation shown here. Also the incorporation of an abrasive improves cleaning power for hard surface cleaners.

Here calcium carbonate, a soft abrasive, is used. This also provides the thickness expected of a cream cleaner. Calcium carbonate is commonly chosen these days due to the popularity of acrylic surfaces that need a soft abrasive to avoid scratching. Sodium carbonate is used here to provide a mild alkalinity to assist the cleaning without being too aggressive on the hands.

Water	42.00
Dodecylbenzene sulfonic acid	2.90
Sodium hydroxide 47%	0.80
Calcium carbonate	51.30
Sodium carbonate	2.00
Coconut diethanolamide	1.00
Perfume, colour, preservative	qs

The sulfonic acid is stirred into the water and when fully dissolved the sodium hydroxide solution is added. This is followed by the other ingredients apart from the calcium carbonate. Once everything in the water is dissolved the abrasive is stirred in. It is to be noted that in this type of preparation sequestering agents must be excluded as they would react with the calcium carbonate.

31 Thick Acid Toilet Cleaner

Strong mineral acids such as the ones used here are very effective at removing lime scale and urinary calculus. Combining the acid mixture with a thickener enables it to cling to vertical surfaces to give prolonged action. In addition, the thickening surfactant improves wetting and provides some detergency action. To achieve this the following formulation uses hydroxyethyl tallow glycinate, an amphoteric surfactant which is especially resistant to strong acid.

Water	77.60
Phosphoric acid 85%	5.90
Hydrochloric acid 37%	13.50
Hydroxyethyl tallow glycinate 45%	3.00
Perfume, colour	qs

The acids are added to the water with stirring followed by the other ingredients. No preservative is needed at these high levels of acidity.

32 Carpet Shampoo

A simple high foaming formulation uses the ever popular sodium lauryl ether sulfate and coconut diethanolamide. The isopropanol (isopropyl alcohol, IPA, propan-2-ol) helps with heavy grease.

Water	51.3
Sodium lauryl ether sulfate 28%	35.7
Coconut diethanolamide	3.0
Isopropanol	10.0
Perfume, colour, preservative	qs

The surfactants are stirred into the water, with warming if necessary, followed by addition of the other ingredients.

33   Glass Cleaner

In addition to the detergent action of the surfactant the cleaning action of this preparation is enhanced by the alkalinity from the phosphate and the grease cutting action of the alcohol. A sequestering agent, EDTA, ensures a sparkling clean surface by chelating metals in the water and on the glass surface, keeping them in solution and ensuring no dried-on dulling effect after rinsing.

Water	92.20
Sodium C13/15 alcohol ether sulfate 27%	3.70
Tetrasodium pyrophosphate	0.60
EDTA tetrasodium	0.50
Isopropanol	3.00
Perfume, colour, preservative	qs

The EDTA and pyrophosphate are mixed into the water and when completely dissolved the other ingredients are blended in.

34   Car Shampoo Concentrate

A very simple preparation combining the powerful detergent action of the anionic dodecylbenzene sulfonate salt with a foaming stabilizer is this car shampoo:

Water	61.0
Sodium dodecylbenzene sulfonate 60%	35.0
Coconut diethanolamide	4.0
Colour, preservative	qs

The anionic surfactant is dissolved in the water followed by blending in of the coconut diethanolamide; warming may assist the dissolving.

35   Screenwash

A large proportion of alcohol is included here, mainly to depress the freezing point of the diluted product but it also assists cleaning of the windscreen and suppresses foam formation. Many surfactants will do the job but the anionic, sodium lauroyl sarcosinate, used here is popular because it is non-smearing.

Water	74.0
Sodium lauroyl sarcosinate 30%	1.0

Isopropanol   25.0
Colour   qs

The ingredients are simply added to the water and stirred at room temperature.

36   Rust Remover

Preparations for the removal of rust from iron and steel typically contain phosphoric acid that converts the insoluble iron oxide (rust) into soluble iron phosphate which can be rinsed away. The presence of a water miscible solvent and surfactant for wetting and detergency are often included as in the following example.

Water   57.6
Phosphoric acid 85%   30.0
Dipropylene glycol monomethyl ether   12.0
Octyl phenol ethoxylate (9EO)   0.20

The acid is added to the water with stirring and then the surfactant and solvent blended in.

37   Thickened Peroxide Bleach

Sodium hypochlorite is the traditional bleach but its use is now being challenged due to environmental concerns promoting a non-chlorine approach. Hydrogen peroxide has a similar bleaching action but fairs better from an environmental point of view and is likely to figure more in formulations of the future. To thicken the peroxide this formulation uses an amphoteric surfactant along with a small amount of sodium chloride.

Water   76.5
Hydrogen peroxide 30%   20.0
2-Hydroxyethyl tallow glycinate   2.5
Sodium chloride   1.0

The sodium chloride is dissolved in a small amount of the water. The peroxide is mixed into the water with stirring followed by the surfactant and the sodium chloride solution.

38   General Hard Surface Cleaner with Pine Oil

In this formulation an alcohol ethoxylate is the main surfactant with its role supported by the soap potassium oleate that is formed *in situ*. The presence of alkali and water miscible solvent also adds to the cleaning action whilst pine oil provides some bactericidal action and gives the familiar pine odour which is associated with cleanliness.

Water	84.5
Oleic acid	3.0
Potassium hydroxide 50%	2.0
Propylene glycol monomethyl ether	3.5
Pine oil	0.5
Sodium metasilicate pentahydrate	1.5
Tridecyl alcohol ethoxylate (7EO)	5.0

The sodium metasilicate is dissolved in the water and then the alcohol ethoxylate stirred in. Oleic acid is blended with the potassium hydroxide and warmed to 50 °C followed by thorough mixing and then addition of the solvent and the pine oil. Finally the aqueous phase is added to the solvent phase and mixing continued until a clear product results.

39   Surface Cleaner/Biocide

Here benzalkonium chloride, a quaternary cationic surfactant, is used for its bactericidal properties and combined with the cleaning action of the alkali and alcohol ethoxylate. Preparations such as these may be diluted to give a wipe-over cleaner and sanitizer that is thin enough to go into a trigger spray bottle. It is important in preparations of this kind to be aware of legislation relating to biocidal chemicals and their applications and to keep the amount of non-ionic surfactant to a minimum so as to prevent the cationic surfactant's activity being inhibited due to its incorporation in the non-ionic's micelle structure.

Water	83.0
Benzalkonium chloride solution 50%	5.0
Alcohol ethoxylate C13/15 (7EO)	2.0
Tetrapotassium polyphosphate 50%	8.0
Sodium metasilicate pentahydrate	2.0

The sodium metasilicate is first dissolved in the water followed by addition of the other ingredients.

40   Hand Gel Kerosene Base

The monoisopropylamine salt of dodecylbenzene sulfonic acid is a good emulsifier for aliphatic solvents. In this formulation it is used in combination with other surfactants, one of which is the soap formed by reaction of the base, triethanolamine, with the fatty acids present in tall oil.

Water	58.0
Odourless kerosene	25.0
Monoisopropylamine dodecylbenzene sulfonate	8.0
Nonylphenol ethoxylate (6EO)	2.0
Tall oil fatty acid	5.0
Triethanolamine	2.0

The water and triethanolamine are mixed. The other ingredients, except for the kerosene, are blended and warmed to 60 °C and the water/triethanolamine solution slowly added with stirring. Finally the kerosene is blended in and the whole allowed to cool.

By now it will be appreciated that it is not only essential to have the correct ingredients in the right proportions but that, also, certain procedures have to be adopted for the mixing. In addition to these it should also be clear that an understanding of the chemistry involved is essential.

There will be some readers who are setting out to prepare their own formulations and it is therefore appropriate at this stage to add a few comments on some important blending considerations.

BLENDING THE CHEMICALS

It is beyond the scope of this book to go into the technology behind chemical blending save to point out that in many cases it is not simply a matter of measuring out the ingredients and stirring them up as one might be tempted to do after a simplistic look at the formulation.

In all the formulations given in this book the amounts of chemicals are quoted in %w/w (grams of substance in 100 g preparation); this is common practise although some liquid preparations use %w/v (grams of substance in 100 ml preparation). Of course where dilute aqueous solutions are involved the density is very near to that of pure water it makes little difference which is used.

But, say, for strong acids and alkalis where the densities are significant then some vigilance is required over whether the percentage is weight or volume based; failure to spot the difference can present a serious problem. The use of percentage amounts is in contrast with more formal chemistry where stoichiometry is important and weights are derived from the number of moles of each substance in a balanced equation.

In formulation chemistry we are, on the whole, not dealing with chemical reactions and reacting masses. However, there are exceptions and, in the planning of some formulations, balanced equations and mole calculations are required as, for instance, where actual chemical reactions occur such as in acid/base neutralization processes.

Without proper planning, and that requires a good deal of chemistry and physics, a formulation may well be doomed. Some preparations are simply solutions and easily prepared; some are dispersions such as emulsions and dispersed particulates in which there is a continuous phase and a disperse phase; some are powder blends.

The more involved formulations such as wax emulsions can demand some strict methodology and without it one could stir the mixture *ad infinitum* without ever approaching the desired outcome. Producing emulsions often requires special equipment such as high shear blenders and mixing vessels with heating and cooling facilities.

The importance of how to mix certain chemicals is learnt early in chemistry studies by considering what happens when concentrated sulfuric acid needs to be diluted. Never add water to acid is the rule. The explanation is that the combination is highly exothermic and the acid is much denser than the water. If water were added to the acid it there would be a tendency for it to float and become hot enough to boil; the outcome would be boiling sulfuric acid ejected from the vessel.

Knowledge of the properties of sulfuric acid and water would enable one to think through the exercise and do things safely even if the rule was not known. And so it is with many mixing procedures although none encountered in the formulation work considered here are as dangerous as the sulfuric acid example.

The compatibility of formulation ingredients needs to be assessed, for example if a cationic surfactant is mixed in with an anionic one the result may be a fatty deposit that creams out of solution, refuses to re-dissolve and is inactive as surfactant. Although most formulation work is preparation of mixtures it must not be assumed that chemical reactions do not occur.

Making small scale preparations in the laboratory and studying a variety of mixing procedures and order of additions is usual before any larger scale work is undertaken. The following list outlines some of the more common aspects of chemistry and physics encountered in formulation blending.

An appreciation of the science of mixing is important for the formulation chemist if he/she is to avoid hazardous and expensive situations arising. It is not enough to search around in the hope of finding a set of rules on what to do. Of course some rules, like the sulfuric acid one, exist and should be taken as guidance but many formulations are too complex for generalizations and the formulation chemist often has to work out procedures for himself/herself.

When there is a need to mix a surfactant concentrate with water the surfactant must be added to the water. To do it the other way round may well result in a thick gel (see micelle structures and viscosity section) that is difficult to dissolve.

Addition of salts (sodium chloride, sodium silicate *etc.*) to surfactants and their solutions is best carried out by first dissolving the solid into a small amount of the water so that a solution is mixed into the surfactant instead of a solid. If the solid were mixed in directly there is the risk of localized gelling around the solid particles as a result of the electrolyte effect on the surfactant. Gel coated solid particles may then be a problem to disperse and dissolve. Behaving like a slippery bar of soap such a particle defies all attempts to break it up or separate it from the remainder of the mixture. For it to dissolve has become a diffusion controlled process in which it has to be left, maybe for some days, for the water molecules to diffuse through the gel matrix, dilute it and reach the dry powder at the nucleus of the offending particle. As the rate of diffusion is proportional to temperature then warming up the blend may hasten the process.

Preparations involving the use of waxes or thick oils should utilize the hot mix method in which the oil phase and the aqueous phase are heated separately to 60–70 °C before combining. It is usually better to have the water slightly hotter to avoid cooling of the oil on mixing as this may result in some solid 'freezing' out of the liquid or sudden thickening that then jeopardizes the emulsification process.

Heating of oils and aqueous solutions should always be by indirect means such as a steam bath. To heat waxes and oils on direct heat can lead to a dangerous superheated layer at the bottom of the vessel and even thermal decomposition. Where heat is to be applied to enable blending it is important to ensure that certain of the substances are protected from too high a temperature. For example, pearlizing agents put into shampoos and bath products work by having minute plate-like crystals that scatter the light and produce the pearl effect. If these get too hot they soften and their well-defined shape is lost, as is their effect.

Thickening agents supplied in powder form, such as hydroxyethyl-cellulose, can be slow and difficult to disperse and dissolve. Again it is the localized gel effect that is the culprit. Thinking through the chemistry of what happens should lead one to conclude that if the powder is first dispersed into a non-aqueous liquid, in which it will not thicken, it can then be added to the aqueous blend without problem. Ethanol or glycerol (propane-1,2,3-triol) are often used for this purpose. Where there is the option of high speed stirrer the powder may be sprinkled into the vortex but then there is the risk of aeration and time wasted waiting for the bubbles to disappear.

Care needs to be taken when stirring surfactant preparations to avoid entrapping air. This is particularly important where thick preparations are being used as it may be near impossible to remove the air bubbles. Certain additions and adjustments, for instance fragrance addition and pH adjustment, are often left to the final stages. With regard to pH it is wise to remember that the value may well continue to change for quite some time in the final blend – when everything else seems stable the pH can be going on an excursion.

MODIFYING FORMULATIONS

In many of the formulations discussed it is seen that the major component is water and we might be tempted to conclude that a better product would result if less water were put in. Water plays a vital role in most formulations. It has to maintain all the components in solution or colloidal state and act as vehicle to deliver all those components to the site of action.

Too little water, or looking at it the other way, too much ingredient, often results in dissolved salts precipitating out or surfactant kicking out, creaming or forming the liquid crystal phases referred to earlier. This is common in formulation work even after much effort in the planning stage.

The following scenario is not too far fetched. A novice formulation chemist, in his enthusiasm to produce a high quality, high performance superb value product made it too concentrated. In appearance the product seemed fine at first and it worked a treat, maybe better than any other product on the market. But overnight, when the temperature in the laboratory dropped, a thick sediment built up at the bottom of the container, due to precipitation, and, worse, it resisted all manner of efforts to re-dissolve it.

Thinking that he had incorrectly weighed one of the chemicals the formulator prepared another sample according to the same formulation. This time he happened to keep it in a warm place overnight. Inspection the next day was accompanied by shouts of delight as the preparation was completely free from sediment. But, on closer scrutiny, it was noticed that a creamy layer floated on the surface. The high concentration of surfactant and warm conditions had caused it to kick out. So back to the drawing board.

After much painstaking work the novice discovered that the only way to get his preparation to stay together was to water it down. This he did and the result was a good product but it was not the super performance one that he had envisaged driving his competitors out of the market with. In fact it was very similar to what other manufacturers were selling. However, the lesson was learnt, for that preparation at least: we must have a sufficiently large proportion of water to hold the whole thing together.

STABILITY

The above scenario brings us on to stability. Many surfactant preparations are energetically (thermodynamically) unstable and will always be ready to break down. Some do not break down because they have kinetic stability. Like a mixture of petrol and air, the mixture has the potential to explode because it is energetically unstable but it will remain forever as the mixture because it is kinetically stable. However, if an outside influence, in this example a spark will do, disturbs the situation then the system rearranges itself to a more stable situation by losing energy. An explosion occurs and changes the petrol and air – hydrocarbon and oxygen – into carbon dioxide and water.

So in formulation work it has to be accepted that the preparation has the potential to break down. The hope is that it has enough kinetic stability to have a reasonable shelf life. If it can get to the age of two before having a break down then that is usually acceptable. The skill in formulation work is to increase kinetic stability as was discussed in the section on emulsions

Once an effective and apparently stable preparation has been obtained, and the primary objective fulfilled, the next stage must be to assess long-term stability and see what happens when the preparation is subjected to the repeated temperature fluctuations, thermal cycles, it will experience during its life. Sadly it often happens that one's pride and joy emulsion turns out to have a lifetime of but a few days once exposed to the rigours of thermal change. Temperature fluctuations, particularly during transport and storage can subject the product to anywhere between sub-zero levels up to around 40 °C. Thus an important element of product testing is thermal cycling.

QUALITY OF CHEMICALS

The quality of raw materials must be taken into consideration. For a shampoo chemicals that are of the toiletries and cosmetics grade must be used and not the general commercial grades that one would choose for, say, an industrial degreasing preparation. Furthermore it is a prudent measure to always be aware of legislation and it's most recent changes relating to preparations offered for sale. Thus the publication of any EEC Directives, examples: Cosmetics Directive, Biocidal Products Directive, relating to various chemical preparations need to be watched. Other important information to be taken into consideration in the planning stages is that from the Material Safety Data Sheets (MSDSs) for the proposed raw materials; here will be found essentials on CoSHH and environmental matters.

ATTEMPTING A FORMULATION

Anyone with only the vaguest idea of formulations can get some raw material sample from suppliers, chuck them together and end up with a working formulation. But to put together a preparation that offers maximum performance, is cost effective, stable and can hold its own in a highly competitive market place requires good technical knowledge and some serious planning.

It may well turn out that one of the listed formulations seems to be effective at achieving its intended purpose but this must not be taken as an indication as to its fitness as a marketable preparation. This should be done only after some fine-tuning and a good measure of testing. The formulations reported above should be treated quite simply as starting points or as indicators of the types of chemicals involved in formulation work.

It is essential, if a preparation is to be a marketable product, that all the aspects are understood. In addition to those few items considered above there are related considerations: safety, directions and restrictions for use, cost, shelf life (preparation and container), disposal and environment, packaging and the interaction of the preparation with that packaging.

Chapter 5

Environment and Resources

Increasingly these days it is important to understand resources, sustainability and environmental burden when looking at applications of chemicals. Consumers are becoming more aware and are making demands that force products to be more environmentally friendly. In response to this many industries have adopted environmental management procedures in which their products and manufacturing processes are examined with a view to becoming 'greener'.

As such it is wise to stand back and take an overview of the role that a formulated product may play and the unavoidable effects that will result from a product's existence. An important approach to this is to make a 'cradle-to-grave' assessment and see what demands are put on resources and the environment by the manufacture, use and disposal of the product. Other similar assessments, Life Cycle Assessment (LCA) and Life Cycle Inventory (LCI), are also used but the principle is the same.

To carry out a 'cradle-to-grave' assessment requires a good degree of thoroughness and an open mind for there is much complexity and many surprises. A product that appears at the planning stage to be environment friendly and using renewable resources may turn out to score less than one based on a non-renewable resource such as petroleum.

Unfortunately, as manufacturers are keen to move towards eco-labelling of their products there has been a tendency to steer a path in a favourable direction and this has done nothing for the reputation of such assessments. This is an area rife with creative accounting. However, 'cradle-to-grave' assessment is a relatively new area that has developed since the 1970s and a lot of refinement needs to be carried out such as in the methodology.

Over recent years there has been a tendency to some very simplistic thinking on the lines that if a product is made of natural materials then it is intrinsically friendly to the environment. This generalization has also picked up some more woolly thinking on the way, for example, natural materials are non-toxic and are biodegradable. If only things were so simple.

But the reality is complex and requires more than sweeping generalizations. The sections that follow are simply examples of what needs to be taken into consideration in any 'cradle-to-grave' assessment to see how a formulated product fairs. In assessments of this nature there is always the problem of defining the beginnings of the product.

Take, for example, a simple dish-wash liquid and examine it being 'created' in a stainless steel mixing tank in a chemical works. Do we start the exercise with an examination of what is consumed in its making: water, surfactants, salt and electricity? Or do we take into consideration the stainless steel tank which was made specially for this job. And what of the concrete base upon which the tank is bolted and the electric stirrer with its motor containing lacquered copper wire, PVC insulation in an iron case coated with cellulose paint which contained VOCs? Where do we start?

It might be argued that the mixing tank and the motor have a long life and as such have a small environmental burden in comparison with the thousands of tonnes of dish-wash liquid that are mixed there and the megawatt-hours of electricity consumed in the same amount of time. So there is some justification for starting the analysis at the mixing stage and going on from an examination of the water, surfactant, salt and electricity.

Thus the mixing tank *etc.* are omitted and the focus is on the chemicals and energy. Even in the crudest of assessments we need therefore to get data on these and this means a study of how they are made. In the following sections the manufacture of the materials required for chemical preparations is outlined but details of the energy part of the equation are beyond the scope of this book.

MANUFACTURE AND RESOURCES

This is confined to the main chemicals used for preparing the types of formulations covered in this book: surfactants, solvents, chelating agents and some of the salts are included. This may seem a rather narrow spectrum of chemicals but there is plenty of potential in just these few examples for anyone choosing to dig deeper.

Surfactants

We have seen that surfactants consist of two essential parts, the lipophile and the hydrophile. Thinking of surfactants in this manner is not only convenient to understanding how they work but is also helpful in looking at their manufacture. Lipophiles and hydrophiles, in general, come from quite separate directions and are then joined to form the surfactant molecules. An exception here is the ethylene oxide/propylene oxide (EO/PO) block polymers discussed earlier. With these surfactants both lipophile and hydrophile are made in similar chemical processes.

Lipophiles

The requirement of the lipophile is that it shall be a long chain non-polar group. This need is met by the hydrocarbon chains found in natural fatty and oily substances. Many of these are available in natural sources, animal fats or plant oils, and the chain lengths of these nature-made molecules are ideal for most surfactant requirements. Natural lipophiles from plants are not pure substances but have a distribution of hydrocarbon chain lengths. They are frequently used as such but for certain applications they may be partially purified to narrow the distribution.

In addition to a mixture of chain lengths these plant products also contain saturated and unsaturated molecules. Thus we get the following alkyl mixtures as lipophiles: cocoyl, lauryl, myristyl, palmityl, stearyl, tallowyl. So here there seems to be plenty of scope for the renewable resource path. In some instances hydrogenation of the double bond in the unsaturated molecules is carried out but this then puts reliance upon petroleum for the hydrogen and energy input.

Opting for plant sources, and to a lesser extent animal sources, of lipophiles avoids to some extent the atmospheric CO_2 problem. As the plants grow they take up CO_2 then we use the plants for our lipophiles, or whatever, and the carbon is ultimately returned to the atmosphere as CO_2. There is no net change: one molecule of CO_2 is simply borrowed from the atmosphere and returned some time later. In this respect plant sources seem ideal but the lipophiles from plants are not just there for the picking; they are chemically combined as glycerides and have to undergo chemical reactions to obtain them.

This is demonstrated by the soap making reaction considered in the surfactants section. Extracting palm kernel oil, one of the major routes to the C_{12}–C_{14} fatty alcohols, from its plant source does not involve chemical reaction, but turning the oil, an ester, into fatty acids or fatty alcohols does. So other chemicals may come into the equation. In the basic process of soap making sodium hydroxide, or similar alkali, is required. And, in turn, that has to be manufactured in chemical reactions; you can't just dig it up out of the ground – which would be non-renewable anyway.

Whichever way the plant fatty acids are to be used as lipophiles there is the problem of other chemicals required for the process. In addition there is the energy input for extraction, reaction and a host of related activities such as transport. Furthermore the plants are likely to be grown in soil artificially nourished by fertilizers that are made from Haber ammonia (see later).

The equations are complex and what at first sight appears attractive loses some appeal after an in depth examination. Because of the present chemophobia many people are drawn into romantic views of being in harmony with nature, making everything from plants that can be grown in their own gardens and not harming the environment. The problem with this is that it fails to take account of the fact that we live highly synthetic lifestyles. The substances we want such as soaps and detergents are not a part of nature. To make them and use them is not natural and it results in environmental damage. To be realistic we have to accept this; to be responsible we have to minimize it.

Lipophiles are also made from petroleum and two routes are commonly used: Ziegler synthesis in which aluminium is used as a catalyst in the stereospecific polymerziation of ethene and the oxo synthesis (the hydroformylation reaction) where an alkene is reacted with carbon monoxide and hydrogen over a cobalt catalyst. These are shown in Figure 5.1. Thus in a wide range of synthetic lipophiles from petroleum there are some aromatic members such the alkyl phenols. Plant-derived lipophiles of the unsaturated variety are sometimes used for catalytic hydrogenation to give saturated lipophiles.

Hydrophiles

Although nearly all lipophiles are chemically similar in that they are long-chain hydrocarbon structures the same generalization cannot be applied to the hydrophiles for here there is a variety of types with differing chemistry. Soap is a fatty acid salt in which the carboxyl group is the hydrophile. When soap is made the hydrophile is already a part of the lipophile and no separate process of attaching a hydrophilic head to a lipophilic chain is required. With the synthetic surfactants things are different.

Sulfation and sulfonation were the first routes to synthetic surfactants. The first commercially successful surfactant, Turkey Red Oil, was sulfated castor oil produced by reacting castor oil with sulfuric acid. From then on a whole range of sulfates and sulfonates made from different

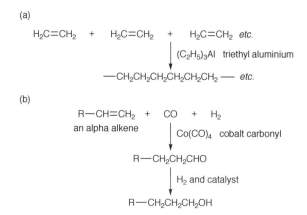

(a)

$$H_2C=CH_2 \quad + \quad H_2C=CH_2 \quad + \quad H_2C=CH_2 \; etc.$$

$(C_2H_5)_3Al$ triethyl aluminium

$$—CH_2CH_2CH_2CH_2CH_2CH_2 — etc.$$

(b)

$$R—CH=CH_2 \quad + \quad CO \quad + \quad H_2$$
an alpha alkene

$Co(CO)_4$ cobalt carbonyl

$$R—CH_2CH_2CHO$$

H_2 and catalyst

$$R—CH_2CH_2CH_2OH$$

Figure 5.1 *Lipophiles can be synthesized from petroleum derived hydrocarbons.* (a) *Ziegler polymerization of ethene using aluminium catalyst produces linear chains of even numbers of carbon atoms.* (b) *Oxo hydroformylation involves reaction of an alpha alkene with carbon monoxide and hydrogen over a cobalt carbonyl catalyst to give chains with even or odd numbers of carbon atoms that are mainly linear but sometimes branched*

lipophiles using sulfur trioxide, oleum ($H_2S_2O_7$) or sulfuric acid were developed.

The essence of the process is the attachment of SO_3 or SO_4 to the hydrocarbon as shown by the examples in Figure 5.2. The distinction between sulfonate an sulfate is a subtle but important one; in the sulfonate the attachment is by means of a carbon–sulfur bond whereas in sulfation, which is the formation of an ester, attachment to the hydrocarbon uses a carbon–oxygen bond.

Neither sulfur trioxide or sulfuric acid are available naturally; they have to be manufactured and that means obtaining raw sulfur from natural deposits, burning it to sulfur dioxide and then catalytically oxidizing the product to the trioxide. A bonus of this reaction is that during the first stage during which the sulfur is burnt much heat is produced. This is not wasted by simply letting it seep away to the environment but is, instead, used to raise steam and generate electricity, a valuable by-product.

Second in importance to sulfonation/sulfation is ethoxylation. This is extensively used to form the hydrophile for many non-ionic surfactants as well as boosting the hydrophilic properties in some anionics, for example, alcohol ether sulfate instead of the alcohol sulfate. To bring about ethoxylation requires ethylene oxide and this is produced by catalytic oxidation of ethylene which, in turn, is made from petroleum.

Ethylene oxide is an unpleasant gas and there is some concern that it

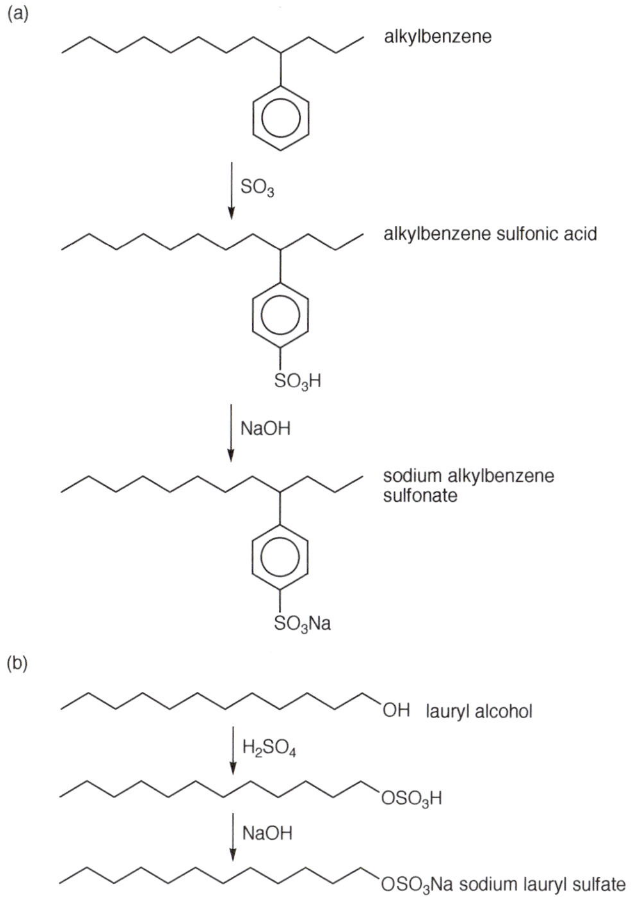

Figure 5.2 *Sulfonation and sulfation to form the hydrophilic head groups. (a) Sulfonation of alkyl benzene to alkylbenzene sulfonic acid followed by neutralization with sodium hydroxide to make the anionic surfactant, sodium alkylbenzene sulfonate. (b) Sulfation of the fatty alcohol, lauryl alcohol, to lauryl hydrogen sulfate followed by neutralization with sodium hydroxide to sodium lauryl sulfate*

may be carcinogenic but that does not mean that an ethoxylate chain behaves likewise. Ethoxylation of fatty alcohols, naturally derived or man-made from petroleum, and alkylphenols accounts for a large proportion of non-ionic surfactant production. Examples of ethoxylation are shown in Figure 5.3. Thus, even where a naturally-derived fatty alcohol is used the non-ionic ethoxylates rely heavily upon petroleum for the ethylene oxide; there is no natural and renewable source of ethylene oxide. The possibility of making it from ethanol which can be made by fermentation of natural carbohydrates exists but such a diverse route is unlikely whilst petroleum-based ethylene oxide is available.

During the ethoxylation reaction it is never possible to ensure the same number of ethylene oxide groups on each molecule of lipophile. The reaction is allowed to continue until the theoretical amount of ethylene oxide has been consumed. A random distribution of products is therefore obtained in which the average number of moles of ethylene oxide per mole of lipophile is used in naming the product. For example, in the making of an alcohol ethoxylate aimed at having six moles of ethylene oxide what may well be obtained is a mixture ranging from zero ethylene oxide content, *i.e.*, unreacted alcohol, to ethoxylate molecules with twelve or even more groups.

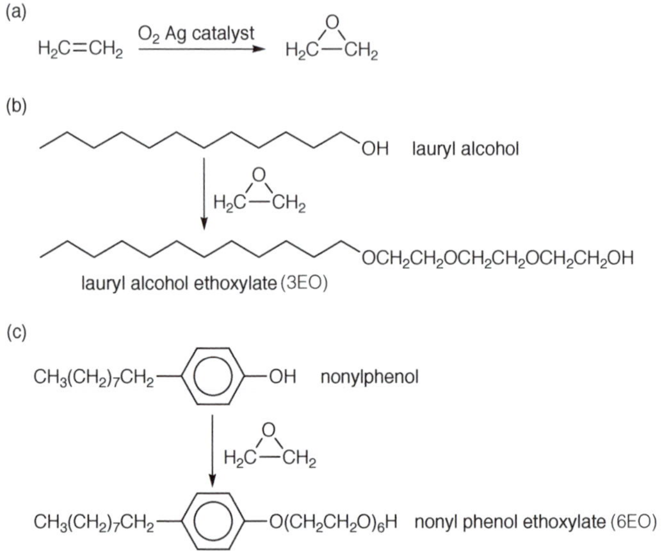

Figure 5.3   *Ethylene oxide and ethoxylation: (a) production of ethylene oxide, (b) ethoxylation of lauryl alcohol to lauryl alcohol ethoxylate 3EO, (c) ethoxylation of nonylphenol to nonylphenol ethoxylate 6EO*

There are few naturally-derived hydrophiles, the main ones of interest for surfactant manufacture are the carbohydrates glucose and sorbitol. Reacting these with appropriate lipophiles gives rise to surfactant molecules such as the alkylpolyglucosides (AGs) and sorbitan fatty esters. The former are currently attracting a good deal of interest for use as as detergents, secondary surfactants, wetting agents and hydrotropes whereas the latter are widely used as emulsifiers especially in personal care products. Figure 5.4 gives a simplified outline of the manufacture of these surfactants.

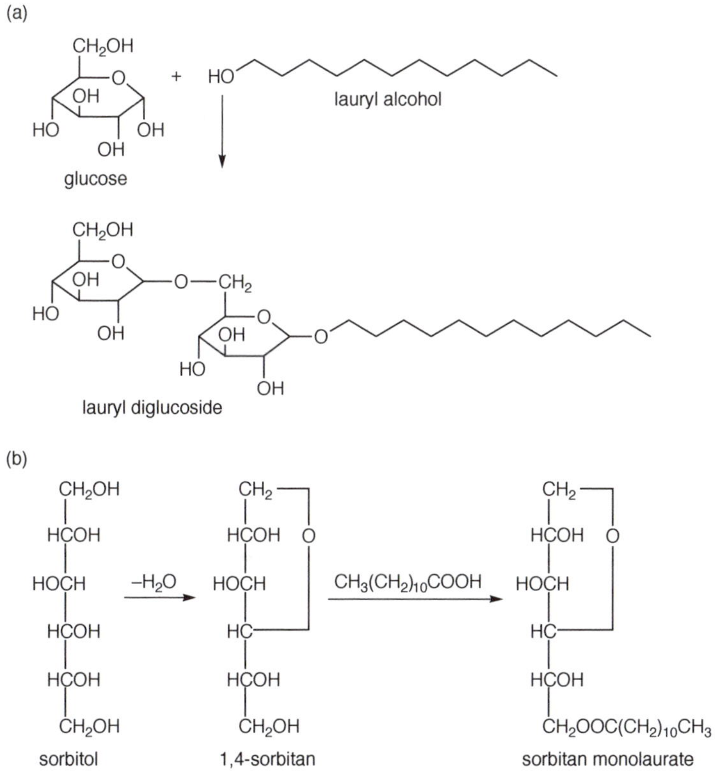

(a)

glucose + lauryl alcohol

lauryl diglucoside

(b)

| sorbitol | 1,4-sorbitan | sorbitan monolaurate |

Figure 5.4 *The carbohydrates glucose and sorbitol can be used as the hydrophiles of surfactants and when these are combined with plant-derived lipophiles the surfactant raw materials are entirely plant based. Simplified reaction scheme for manufacture of* (a) *lauryl diglucoside,* (b) *sorbitan monolaurate*

Where the lipophile attached to the glucose or the sorbitan is a plant fatty alcohol, for example lauryl, the whole of the surfactant molecule is derived from natural renewable substances. The raw materials are in abundance, relatively cheap and are renewable. Furthermore they do not contribute to additional atmospheric CO_2: their production consumes CO_2 in photosynthesis, their destruction releases CO_2; a one to one basis exists. This is a steady state situation and represents sustainable industrial chemistry.

Another attraction of these surfactants is that when they finally go down the drain to the sewage treatment works they are readily biodegraded because they represent what is the regular diet for the bacteria involved in the biodegradation reactions. Alkyl polyglucosides are therefore environmentally benign. Furthermore they have low toxicity and are low on the irritancy scale.

These surfactants are enjoying a healthy growth and this bodes well for the future. This seems like the ideal we have all been looking for but, as with most things that appear to be the perfect solution when first considered, this is not quite utopia. The alkyl polyglucosides are currently difficult to manufacture, expensive and unstable in acid conditions which excludes them from certain formulations.

Also, because of by-product impurities, they are dark brown unless of the bleached variety, which are even more expensive. However, the unbleached ones do find use in industrial surfactant preparations where colour is not a problem. The bleached varieties are used in cosmetic and other personal care preparations. AG technology is developing and glucamines and glucamides are now following on to open up the applications.

Related to these glucose surfactants are the sorbitan fatty esters, for example sorbitan trioleate, that are largely used as emulsifiers in, for example, cosmetic creams. Sorbitan esters are produced from the sugar, sorbitol and fatty alcohol. The amount of hydrophilic character of sorbitan esters can be increased by ethoxylation but this is moving away from the natural aspects towards the synthetic.

Phosphate esters of fatty alcohols are manufactured and add to the range of anionic surfactants. In these the hydrophile is derived from phosphoric acid, and that in turn comes from phosphate rock – see Figure 5.5. To convert the extracted rock, which contains tricalcium phosphate, treatment with sulfuric acid is required and this adds significantly to the non-renewable sources aspect.

$$S \xrightarrow[\text{air burn}]{O_2} SO_2 \xrightarrow[\text{catalyst}]{O_2} SO_3 \xrightarrow{H_2O} H_2SO_4$$

$$Ca_3(PO_4)_2 + 3H_2SO_4 + 6H_2O \longrightarrow 2H_3PO_4 + 3CaSO_4.2H_2O$$

phosphate
rock

Figure 5.5 *Sulfuric acid and phosphoric acid are essential to the current methods for making many surfactants*

To produce the phosphate ester the lipophile, for example lauryl alcohol, is reacted with the phosphoric acid in a condensation reaction. Mono, di and tri-esters are possibilities due to phosphoric acid being tribasic but, in general, most phosphate ester surfactants are the di-esters. The remaining acidic hydrogen is neutralized by sodium hydroxide to give the salt. The reactions are shown in Figure 5.6.

sodium dilauryl phosphate

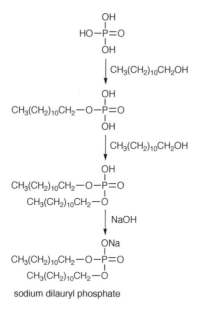

Figure 5.6 *Phosphate esters are formed when a fatty alcohol such as lauryl alcohol undergoes a condensation reaction with phosphoric acid. Most of these types of surfactants are the di-esters with the remaining acidic hydrogen neutralized with sodium hydroxide*

Nitrogen, in the form of ammonia derivatives, also plays a significant role in surfactant chemistry. All the nitrogen compounds involved here rely upon synthetic ammonia from the Haber process (Figure 5.7). This industrial procedure involves a high pressure, high temperature combination of nitrogen and hydrogen in a catalytic reaction. Nitrogen as one reactant is obtained from the air by liquefaction and fractional distillation, a physical extraction process, but the other reactant, hydrogen has to be manufactured and this is usually from petroleum or natural gas.

(a)

$$N_2(g) + 3H_2(g) \rightleftharpoons 2NH_3(g)$$

(b)

$$C_6H_{14}(g) + 6H_2O(g) \rightleftharpoons 6CO(g) + 13H_2(g)$$

$$CH_4(g) + H_2O(g) \rightleftharpoons CO(g) + 3H_2(g)$$

Figure 5.7 (a) *The Haber synthesis of ammonia fixes atmospheric nitrogen in a high pressure, high temperature catalytic reaction.* (b) *Hydrogen for the Haber reaction is obtained from petroleum or natural gas*

Taking all these factors into account it will be immediately appreciated that the Haber process puts a big demand on hydrocarbons from petroleum and natural gas. Furthermore, large amounts of energy are used and this adds to the fossil fuel demand. Although the Haber process does not introduce additional nitrogen into the environment it is increasing the proportion of fixed nitrogen and this may have some environmental consequences. The total amount of nitrogen fixed by the Haber process is now huge; most of it is for making fertilizers.

Figure 5.8 shows examples of the importance of ammonia derivatives in the different categories of surfactant: anionic, cationic, non-ionic and amphoteric.

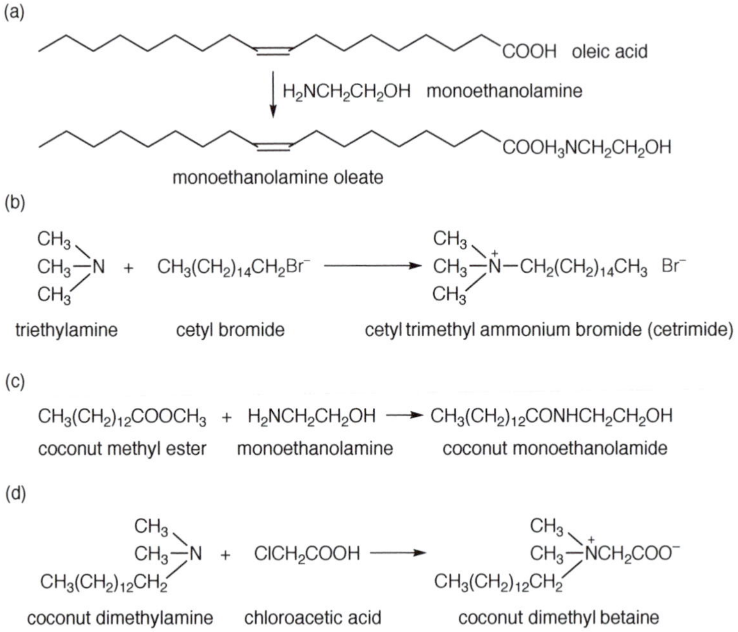

Figure 5.8 *Examples of ammonia derivatives in surfactant manufacture: (a) anionic, monoethanolamine oleate, (b) cationic, cetyl trimethyl ammonium bromide, (c) non-ionic, coconut monoethanolamide, (d) amphoteric, coconut dimethyl betaine*

Solvents

Hydrocarbons from petroleum come top of the list of solvents currently used in many formulations, particularly those that have a degreasing function. These hydrocarbons are also converted into alcohols, glycols, ketones and esters by various reactions in which oxygen, in the form of different chemical groups, is introduced into the molecules.

Where nitrogen derivatives of hydrocarbons are required the nitrogen itself comes from Haber ammonia. Thus, conversion of hydrocarbons into amines, ethanolamines and the like is reliant upon an appreciable energy input. For other hydrocarbons converted into compounds containing phosphorus or sulfur then it can be generally assumed that the synthesis will rely upon phosphoric or sulfuric acid.

Apart from petroleum as a source of hydrocarbons there are plant sources but these are rather limited in comparison to both volume and variety when put alongside the petroleum ones. As is often the case with plant-derived chemicals the means of production is less controllable than for petroleum. However, there is a steady growth in these naturally derived solvents. In addition to traditional plant-based solvents such as turpentine there are some more recent ones. Where plant-based chemicals are used as feedstocks for chemical conversion into different solvents then much of what was noted for the petroleum also applies. Examples of nature-made and man-made solvents appear in the earlier section on solvents.

Halogenated solvents are widely used with the chlorinated ones coming top of the list. For their manufacture hydrocarbons are used and are reacted with chlorine. The chlorine is manufactured from the electrolysis of molten sodium chloride and although the salt is plentiful large amounts of energy in the form electricity are used.

The most widely used solvent in making chemical formulations is one that is completely natural, water. However, drawing water from the mains tap is taking a substance that has been purified by chemical means such as treatment with ferric chloride [iron(III) chloride] to precipitate impurities and subsequent disinfection with chlorine. For a good deal of formulation work further purification is needed. Whatever means is adopted to bring this about it is going to require energy and increase the demand on fossil fuels.

Chelating Agents, Salts, Alkalis

Chelating agents such as NTA and EDTA being nitrogen compounds rely upon Haber ammonia. Phosphates and polyphosphates can be regarded as being similar to the phosphoric acid discussed above. Sodium hydroxide used as an alkali or for making the sodium salt of anionic surfactants is from the electrolysis of aqueous sodium chloride solutions.

Sodium carbonate is generally from the Solvay process in which several reactions are carried out and take as their raw materials sodium chloride, calcium hydroxide (from limestone) and carbon dioxide. Clearly natural, non-renewable, resources are being consumed in all these reactions but the need for restraint does not have the same urgency as with the fossil carbon resources. However, the processes require energy and this means more CO_2 into the atmosphere.

Packaging

From a 'cradle-to-grave' point of view there has to be some account taken of the packaging that the raw material chemicals come in: steel drums, polythene containers, paper sacks *etc.* However, for the chemist involved in planning a formulation it is the packaging aspects of the finished products that are more significant as this will most certainly outweigh that from incoming raw materials side. The packaging of chemical preparations is often a major item, especially in the field of cosmetics and toiletries. Packaging will be looked at again in the examples given below.

ENVIRONMENT AND POLLUTION

Pollution from manufacturing processes and use of chemicals is inevitable. In considering this pollution we need to assess chemicals in terms of Ozone Depletion Potential (ODP), Global Warming Potential (GWP) and biodegradability. Of course the latter is more concerned with chemicals in water and soil whereas the former ones relate to volatile chemicals, mainly Volatile Organic Compounds (VOCs), in the atmosphere.

However, some of the chemicals going into the atmosphere are to be rained out and absorbed by the soil and watercourses. Other considerations may include eutrophication and atmospheric acidification. Biodegradation of chemicals in water requires oxygen and that comes from the dissolved oxygen in the water. It is desirable that the chemicals degrade, ultimately to simple molecules such as CO_2, H_2O, NH_3, SO_2 *etc.*, but whilst that is happening the water is being robbed of dissolved oxygen. This is measurable in the laboratory as Biochemical Oxygen Demand (BOD); total chemical oxidation is determined using Chemical Oxygen Demand (COD).

Chemical Production

Looking at the preparation of chemical formulations there are clearly two areas of pollution: manufacture of the raw materials and the blending of the formulation. In both these areas there is going to be a certain amount of pollution mainly due to small quantities of chemicals getting into the atmosphere, into the ground or into watercourses. Inevitably there will be occasions when large scale pollution incidents occur due to the unexpected. In addition, the energy used in the chemicals production is going to mean more CO_2 into the atmosphere.

Formulated Products in Use

Many of the formulations looked at in this book are for household use; as such they are going to carry out a particular function and when this is completed the chemicals are waste substances and are destined to end up in the environment. A few preparations containing solvents have the potential for an element of recycling but this is only ever realistic in industrial situations where large quantities can be stored up and sent for processing such as distillation to recover the solvent. For most of the products considered here disposal of them is a matter of sending them down the plughole and on their way to the sewage treatment works.

There will inevitably be some of these products where disposal amounts to them soaking into the ground and/or draining into watercourses. This should not happen, but it does and it seems inevitable that it will continue. For example, washing the car for most of us results in the diluted car shampoo going into a ground water drain which then carries it straight to the nearest watercourse. Some of the shampoo will soak into the ground but ultimately that also will end up in a watercourse.

It is important here to recognize that there are often two drainage systems: one is intended for rain water and conveys water straight to the nearest stream or river, the other is for carrying foul water to the sewage works for treatment. Of course not everyone lives in an area where the foul water goes to a sewage works but there is always some form of treatment before the water goes into the river. If this were not problem enough we also have to consider what the used shampoo contains in terms of the dirt it has removed from the car. No longer is it shampoo and water but it will, after having done its job, also contain: waxes and silicones from car polish, metal dust from the brakes (and this contains heavy metals) and traces of oil.

That is quite a chemical cocktail to be sending into the local stream and it is almost certain to have some toxic effect upon the aquatic life. Eventually the organic chemicals will biodegrade, to small and relatively harmless molecules, but in the interim much damage can be done. And the process of biodegradation removes dissolved oxygen from the water.

Even where a chemical is not actually toxic it can still kill aquatic life by, as in the case of a surfactant, altering some physical properties such as surface tension. The heavy metals, some of which are highly toxic, do not degrade and are there forever or they may end up in the food chain. Clearly it is better to take the car to a proper car-wash facility where the dirty water will go to sewage works and be cleaned up before returning to the natural environment.

Biodegradability is an important aspect of aqueous formulations. Although virtually all organic chemicals will biodegrade eventually, what we require are ones that will do so whilst in the sewage treatment stage, before entering the natural environment. Biodegradation involves the progressive breakdown of the organic molecules into smaller molecules and ultimately into carbon dioxide, water, sulfate, phosphate and nitrate *etc.*

Even only partial degradation of an organic molecule before it enters the natural environment is usually better than nothing in that it has probably removed much of the molecule's aquatic toxicity. This is not, however, always the case. Some partial breakdown products may be more of a problem than the original molecule. For example the alkylphenol ethoxylates, as they biodegrade, produce molecules that are thought to behave like hormones and pose a risk to human health.

Toxicity of surfactants in the aquatic environment has been studied for many of them by establishing how they affect algae, daphnia and fish. Modern surfactants, in general, have good biodegradation characteristics but it has not always been like this. In the 1950s the first sodium alkylbenzene sulfonates came onto the domestic market in the form of washing powders and the problem of foaming rivers arrived.

The cause was that the surfactants molecules were slow to biodegrade and passed through sewage treatment works unaffected, taking their detergency properties into the rivers. The surfactants were soon replaced by ones with a different lipophile structure that was quicker to biodegrade.

Chelating agents and how they behave in the aquatic environment are currently the subject of research. Many surfactant formulations contain chelating agents. In some there has been a shift away from phosphates, because of eutrophication worries, towards ethylenediaminetetraacetate (EDTA) and nitrilotriacetate (NTA). This may have helped the eutrophication aspect but unfortunately both these compounds are slow to biodegrade, although it seems that NTA is better than EDTA in this respect.

These chelating agents, once in watercourse, continue to chelate metal ions and move them around thus disturbing the natural balance that is essential to living systems. Chelating agents that have sequestered heavy metal ions whilst in use retain those ions and deposit them in the watercourse when the organic molecule does eventually break down. These chemicals therefore have a capacity to dissolve metal compounds that are normally insoluble in water with the possibility of mobilizing toxic metals into the aquatic environment.

Sediments made up of insoluble compounds of heavy metals are sometimes to be found in river beds. These are not necessarily from man-made pollution as many have natural geological origins. Being insoluble and sitting there as dense sediments they remain inert to living systems but once chelating agents arrive on the scene there is disruption. The metals are leached out and become metal chelates. In this form they are now mobile and toxic to aquatic life. A natural environment that had been stable for maybe hundreds of years and was living in safety with its toxic legacy neatly locked up is then at danger due to the toxic metals being on the loose.

Salts and alkalis will generally remain in solution as their ions and enter the environment as such – there really is nothing to break down. The sodium chloride that was used as the shampoo to thicken it up will end up as sodium chloride in the river and on the way to join oceans of more of the stuff – or to be abstracted downstream to be used as drinking water. By the time it reaches the river it will be extremely dilute and, along with the fact that the sodium chloride is relatively innocuous anyway, so this salt has negligible effect.

We can find similar examples in other products. Machine washing powders often contain sodium sulfate as a filler to give the powder the correct properties. This is also innocuous at high levels of dilution and will have little effect on aquatic environment of a river. Probably the worst that will happen is that some will precipitate out as calcium sulfate due to the ubiquitous calcium ion in natural waters.

However, compare those salts with phosphates and polyphosphates and things appear a little less healthy. Some phosphate will inevitably end up in the rivers although, nowadays, much of it will have been removed during sewage treatment by precipitation as calcium phosphate. A problem here is that although some polyphosphate will break down to phosphate and be precipitated much of it does not; instead it passes through the process as soluble calcium polyphosphates. To make matters worse the polyphosphates are good chelating agents. As such, once in the natural aquatic environment, they can behave in a similar manner to the chelating agents considered above.

The addition of phosphate, a plant nutrient, to aquatic systems is bound to have an effect. Where the concentration is significant and where there is also increased levels of nitrate then there is the potential for eutrophication. However, the regular household and personal care products do not contain nitrate so how can we blame them? The answer here is that the many of these products contain nitrogen-based surfactants: ethanolamides, quaternaries, betaines *etc.* that biodegrade to eventually form the nitrate.

Acidity and alkalinity from household formulations can be quite severe and if allowed to get into the soil or watercourse with only little dilution there would be damaging effects. The problem of acidity or alkalinity is normally countered by high dilution. It is worth reflecting on pH here and just how much dilution is needed to bring a chemical preparation to near neutral.

Take for example a rust remover based on 10% w/w hydrochloric acid. This contains monobasic acid at a concentration of around 3.6% which will have a pH approaching zero. This will increase by 1 for each factor of 10 during the dilution. Thus, if 1 litre of the rust remover is diluted a thousand litres the pH will be about 3. At this pH there would still be serious damage to the aquatic life in a receiving watercourse. A further dilution of the same magnitude would be required to get somewhere near to natural water pH. And that's a lot of water.

VOCs are present in many household formulations as: emulsified insoluble solvents, water soluble solvents, solvents in non-aqueous formulations. In use there will be some loss to the atmosphere depending upon the volatility of the solvent and so GWP and ODP are a consideration. There may be indoor environmental problems such as the degradation of air quality by the VOC and inhalation of this by those who use the preparation. This is particularly so in industrial use where there is the likelihood of repeated and/or long-term exposure to those who use them. Consequently, factors such as occupational exposure limits have to be taken into account.

Most aerosols rely upon VOCs as propellants although a few now use gases such as carbon dioxide or nitrogen. The VOC propellants, for example propane, butane and hydrofluorocarbons, can be present in aerosols in quite large amounts. They are destined, of course, to go straight into the atmosphere when the aerosol is discharged – back to GWP and ODP considerations. Once in the atmosphere these compounds are broken down into simpler molecules, mainly by photolysis and oxidation. The products of these reactions may then be washed out by rain and enter the groundwater.

In Perspective

Clearly many of the everyday formulations impose a burden upon the environment. With regard to the chemical formulations discussed in the foregoing pages that burden is not too heavy and, over the years, where chemicals have been found to be a problem improvements have been made. Furthermore with the trend towards renewable plant-based chemicals things are getting even better.

In many household formulations it is not always the preparation itself that is the problem but often the packaging. This is particularly so with cosmetics. For example a 25 ml jar of hand cream containing 80% water with some oil and emulsifier accounting for the remaining 20% packaged in a jar weighing 200 g does seem to be a slap in the face for the environment: 5 g of active chemical and 200 g of packaging. And the 500 ml dispenser bottle of liquid soap with 120 g of packaging for 100 g of surfactant does look rather wasteful when compared with a bar of soap with a simple paper wrapper which does the same job.

There are many examples which may leave us wondering if the manufacturers of some preparations have any regard for environmental factors or the squandering of valuable resources. However, we must keep things in perspective and judge our chemical formulations and their packaging alongside other environmental burdens.

When compared with the huge amount of damage our beloved cars do to the environment those household chemicals pale into insignificance. Quite simply the drive to the supermarket to buy the goods has done far greater damage than the goods themselves.

Chapter 6

Formulation Analysis

The formulation chemist will, from time to time, be called upon to examine the products from different manufacturers, to assess his/her own preparations and to attend to some basic quality checks on raw materials. When examining the products made by another company it is wise to be mindful of commercial secrecy and the risks involved in acquiring information that is the intellectual property of a third party or which may be protected by patents.

Where analysis of another company's products is embarked upon some serious consideration must be given as to how the results will be used. To use them to produce an exact copy – or as near as is technically possible – for commercial gain may well invite some contentious situations, and to publish the results of such work would certainly be treading on marshy ground. Be that as it may, all commercial formulation work carries with it an element of 'seeing what the competition are up to'.

In the following pages an overview of the analytical and test methods that are commonly used in the formulation laboratory is given. Many of the methods double as investigation tools as well as devices for checking the quality of incoming raw materials and finished products. Although some procedures look to be quite simple none of them should ever be used on a one off basis; in each case experience in using the method and familiarity in the way a particular sample responds is essential. Proper attention should be given to precision, accuracy, reproducibility and repeatability. Planning to ensure a representative sample is obtained is a must. In addition, wherever possible, blanks, standards and known samples should be run in parallel to the test sample.

Many of the methods involve simple procedures and regular laboratory equipment such as analytical balance, oven, furnace, fume cupboard, glassware *etc.* Coupled with good laboratory practise these simple methods give quick, reliable and low cost results. However, a few procedures require expensive instrumentation and, certainly so far as many formulation chemists are concerned, will remain the province of specialist laboratories and research facilities.

None of the methods referred to here assess microbiological aspects of preparations. Such work may well be an essential requirement but is best left to the microbiologists.

The overview that follows is necessarily of a general nature and includes only a brief outline of the more common procedures. Details of the methods listed here – and other methods – are to be found in standard chemical analysis texts and in publications relating to specifications and standards.

For example, the analysis of surfactant products is covered in British Standards (BS): (a) BS 3762, Analysis of formulated detergents, 24 tests described. (b) BS 6829, Analysis of surface active agents (raw materials), 20 test methods. Further methods, some of them overlapping those of the BS, are to be found in other series: International Organization for Standardization (ISO), the American Society for Testing Materials (ASTM), British Pharmacopoeia, Institute of Petroleum and others.

The application of analytical methods for exploring single pure substances may well be reasonably straightforward but when dealing with formulations, which are frequently complex mixtures of substances – and usually impure ones at that, it must be appreciated that one is never playing on level ground.

Interpretation of results obtained from mixtures or components separated from mixtures requires a good degree of caution and some enthusiasm for trying different ways of quantifying a particular component to have any degree of confidence in what is often an ocean of interferences.

Before embarking on any chemical analysis or test it should be borne in mind that mixtures of chemicals, as indeed most preparations are, may well be stable under normal conditions of use and storage but when subjected to the rigours of various laboratory techniques and mixing with further chemicals they may turn into dangerous entities. Take heating for example: a formulation that is seemingly innocuous at ambient temperature may yield toxic fumes or even explode at elevated temperatures.

A risk assessment, however sketchy, must be carried out and include consideration of hazards (inherent properties of chemicals) and risks (probability of a hazard causing harm). Any preparation suspected of containing an oxidizing agent must, of course, be subject to rigorous safety precautions.

The minimum amount of sample should always be used; a few milligrams of substance exploding whilst drying down in the laboratory oven may be no great event but to have several grams do the same thing is an invitation to living dangerously – and forcing the other laboratory staff to do likewise. Performing laboratory tests on unknown mixtures of chemicals is risky enough anyway.

Before starting any laboratory work it should be borne in mind that the results can only ever be as good as the sample and some questions posed: how representative is the sample, who was responsible for sampling, how has it been packaged and stored? A poor sample can waste an awful lot of time and effort.

Even with a good sample and a high level of confidence in the results the interpretation calls for some caution. Quite simply what comes out may not be what went in. Even preparations where no reaction between ingredients was envisaged can throw up a few surprises.

BASIC METHODS

1 Preliminary Examination

The application of the human senses is always the first step; thus, appearance, smell, texture provide the first clues but, of course, the latter two require an element of caution in giving due consideration to a Material Safety Data Sheet, hazard labels and the like. This examination may be extended to good effect by examination under the microscope.

Also at this stage some simple test tube observations can produce valuable clues, such as solubility/dispersibility in water and what happens on acidification or rendering alkaline. For example a foamy solution that on acidification loses its capacity to foam and produces a greasy layer that re-dissolves on addition of alkali often indicates the presence of a soap.

Cautiously heating a small amount of sample is often fruitful: different effects at different levels of temperature, right up to pyrolysis and burning off, are often found. For example, the formation of glassy beads or glaze-like film when a detergent formulation is burnt down at red heat on a spatula over a Bunsen flame suggests it may be wise to later prepare the sample for a phosphate test.

2 Total Solids Content

Where a preparation is believed to contain volatiles such as water or solvents it is useful to determine the amount of dissolved matter. The procedure involves heating a small amount of sample and quantifying the loss by accurate weighing until constant weight is obtained. The method is also of value when dealing with solids/powders for moisture content or water of crystallization.

It is usual to begin the heating by means of a steam bath with fume extraction; this is particularly so where flammable, corrosive or toxic vapours are likely. Final drying is carried out in the laboratory oven; the normal temperature being 110 °C. The remaining material is often referred to as dry matter or solids even though it may be an oily liquid.

3   Ash Determination

The preferred method for ash determination involves the use of a platinum crucible for both the ashing process and any subsequent acid treatment. Frequently carried out on the dry matter from the total solids content the ash determination gives an indication of inorganic components such as salts that are left behind after the organics have burnt off.

The procedure involves quantifying the mass loss, to constant weight, when the sample or the dry matter from total solids determination is heated at 550 °C to burn off all organic matter. Some materials require the addition of oxidants or the application of higher temperatures to remove the last traces of carbon. Converting the ash into its sulfate form is a frequently used method and is effected by treatment of the crude ash with sulfuric acid followed by further ashing. It is common practice to retain the ash for further work.

4   Suspended Solids

This test applies to preparations based on solid suspensions such as fine powders held in, usually, aqueous medium. It involves filtering out, using vacuum filtration, the suspended matter. This is followed by rinsing to remove any soluble matter, drying at 110 °C and weighing to constant weight. Typically a Whatman GF/C glass fibre filter is used in conjunction with a Hartley funnel and the suction provided to a Buchner funnel by means of a vacuum water pump. Samples are sometimes difficult to filter and require pH adjustment to bring about coagulation of the suspended matter that then filters rapidly.

5   Dissolved Solids

This applies to samples containing suspended matter and is usually carried out by using the procedure for total solids to the filtrate after the suspended solids test. Alternatively, dissolved solids may be obtained from the numerical difference between total solids and suspended solids. It should be noted that dissolved solids may also include involatile colloidal matter.

The residual solids from evaporation, or any surplus filtrate, should be retained for further work.

6   Emulsions, o/w or w/o

A few methods are available to establish emulsion type.

(a) In general an emulsion may be readily diluted by addition of its continuous phase; thus, an oil-in-water emulsion, o/w, is dilutable with water whereas a water-in-oil emulsion, w/o, is diluted by oil.

(b) A mixture of two different coloured dyes, in which one is water soluble and the other is oil soluble, may be used. A convenient way is to prepare a mixture of the two dyes in powder form in an inert diluent powder such as alumina. The sample is spread onto a white tile and lightly sprinkled with the dye mixture. Rapid development of blue colour indicates an o/w emulsion or development of red colour indicates w/o emulsion.

(c) Conductivity measurements are also useful. Oil in water emulsions give measurable conductivities that are enhanced by addition of small amounts of powdered sodium chloride as electrolyte – so long as this does not disrupt the emulsion itself. Water in oil emulsions have zero conductivity and show no response to the electrolyte addition.

(d) Cobalt chloride moisture test papers – if pre-dried in an oven and desiccator – when spotted with o/w emulsion undergo instant change from blue to pink but will show no colour response when the emulsion is of the w/o type – prolonged contact may produce a weak change.

7   Solvent Extraction

Applicable to aqueous formulations the principle of this method is to use a solvent that is immiscible with the sample and which is a good solvent for the substance to be separated. Common solvents used here are petroleum spirit 40–60, diethyl ether (ethoxyethane), chloroform (trichloromethane), ethanol. Chloroform is frequently used for the extraction of anionic surfactants from aqueous solutions as in the methylene blue test, which may also be used for quantitative determination, described below. An ethanol extract is particularly useful in identification of surfactants.

In general the extracted matter is recovered by evaporation of the solvent and the extract may then be used for further work. Choice of solvent may be a matter of trial and error but knowledge of how solvents work will keep to a minimum the number of trials. Other conditions, such as pH, are varied to optimize the effectiveness of the extraction process.

8   Soxhlet Extraction

This procedure is confined to solids therefore its application in the analysis of aqueous preparations calls for separation of dissolved/dispersed solids by evaporation and drying off. The dry solid, placed in a porous thimble – usually porous pot – becomes immersed in the solvent as it distils from the flask, condenses and drips into the thimble standing in the extraction chamber.

When the latter is full, siphoning occurs and the solvent plus extracted matter is flushed out and into the flask whereupon the whole process is repeated. By this means the sample being extracted is constantly supplied with fresh solvent. The method is one of hot extraction due to heat received from beneath the extraction chamber rather than any heat provided by the downpour of solvent.

9   Methylene Blue

Used to detect anionics this test is among the more common for quantifying the concentrations of anionics (excluding soaps and sarcosinates), for example, in effluents. Into the dilute sample is added acid methylene blue. This is followed by chloroform extraction during which the anionic–methylene blue complex is carried into the chloroform. For quantitative work the intensity of the blue extract is measured in a visible spectrophotometer along with calibration standards.

10   Dimidium Bromide

Dimidium bromide–disulfone blue mixed indicator is added to acid and alkaline solutions of a surfactant in water and extracted with chloroform into which a coloured complex migrates. The actual colour is indicative of the type of surfactant and by careful application this test can be quite specific.

11   Conductivity

The electrical conductivity between platinum electrodes connected to an AC voltage at 1000 Hz is useful in quantifying the level of ionized compounds in an aqueous solution. In the absence of interfering electrolytes conductivity enables anionic and cationic surfactants to be distinguished from non-ionics, and oil-in-water emulsions from water-in-oil ones.

In general, for dissolved salts, the numerical value of conductivity in $\mu S\ cm^{-1}$ roughly parallels the value for the concentration of dissolved solids in $mg\ l^{-1}$. For precise work, calibration with potassium chloride solutions is usual along with appropriate standards and specified temperature. It should be borne in mind that tap water, which is sometimes used in lower quality preparations, has significant conductivity and the harder the water the greater the figure.

12 pH Determination

pH measures the acidity/alkalinity of dilute aqueous solutions and is usually determined by means of a pH meter with combination electrode. Calibration is carried out against buffer solutions, typically at pH values of 4, 7 and 9, at 20 °C. Being a logarithmic scale means that a change by a factor of ten in the acid/alkali strength brings about a change of one unit on the pH scale. The limits of the scale are represented by solutions of hydrogen ion concentration: at $0.1\ mol\ l^{-1}$ the pH is 1 and for solutions of similar hydroxide concentration the pH is 14.

Thus any pH determination carried out on a strong solution, as indeed most solutions in formulation work are, should be treated with caution as the following example shows.

Example: A caustic detergent blend used for industrial degreasing has a pH of 11.8 but when diluted shows some interesting pH readings: 10% pH = 13.3, 1% pH = 12.9, 0.1% pH = 12.0, 0.01% pH = 10.8, 0.001% pH = 9.6.

It is common practice, when dealing with strong solutions, for pH to be quoted at a certain dilution rather than as 'crude' pH.

13 Acid/Base Titration

Determination of the concentration of acid or alkali by titration is a common method that offers good levels of accuracy and is applicable to solutions and readily soluble solids. Insoluble solids may also be titrated by a modification of this method, see below for an outline of back titration.

The usual procedure involves the quantitative addition of a standard acid by means of burette into a known quantity of base until neutralization is accomplished as shown by colour indicator or pH determination. Where the latter is applied and a pH plot carried out this may reveal more information.

Simple titrations may be complicated by the presence of other components in the mixture. For example, the titration of sodium hydroxide solution often has complications due to the presence of carbonate that has formed on exposure to the atmosphere.

A common application of acid/base titration is in determination of the acid value and saponification number of fats, oils and waxes.

14   Back Titration

This is applicable to certain insoluble components, for example powders and suspensions, and involves reacting the sample with a known volume of standard acid or alkali and then determining by titration how much of the standard has been used up in reaction.

15   Active Chlorine Titration

Normally applied to sodium hypochlorite solutions to determine the available chlorine this test relies upon the redox reaction between iodine, liberated during reaction with added potassium iodide, and thiosulfate ion. This procedure is used frequently when dealing with sodium hypochlorite solutions because they are unstable and even a small temperature increment can result in significant drop in active chlorine level.

16   Hydrogen Peroxide Titration

Another redox reaction utilizing the iodine/thiosulfate reaction is the determination of peroxide content. Here a sample is treated with excess potassium iodide, which is oxidized by the peroxide to liberate iodine. The iodine is determined by titration with standard sodium thiosulfate solution using starch to improve the end point. Modifications for different sample types include the use of a catalyst and a water immiscible solvent to dissolve the iodine and remove it from the reactive aqueous phase.

17   Iodine Value

In addition to its use in assessing the content of unsaturated matter in vegetable oils and fats the iodine value is also useful in the identification process. The method, which involves determining the quantity of iodine that reacts with the carbon–carbon double bonds in the substance, is carried out in a specially designed flask which prevents access of oxygen during the reaction stage.

It is usual to use Wijs solution which contains iodine monochloride in glacial acetic acid as the source of reactive iodine. The reaction mixture must be carried out in the dark as far as possible. At the end of the reaction, potassium iodide is added and the liberated iodine titrated with standard sodium thiosulfate solution. The value obtained, the iodine value or number, is the number of grams of iodine absorbed by 100 g substance.

18   Tests on Ash

Ash is the inorganic matter left after all the organics have been completely oxidized. Where the ash can be successfully brought into solution, normally by means of acid digestion, it is a valuable aid in that the solution may then go for instrumental analyses such as Flame Photometry or Atomic Absorption Spectrophotometry for metal and Ion Chromatography for the anions.

As a preliminary to this acid digestion and instrumental analysis the ash itself may be used for some simple tests such as spot tests for phosphate, sulfate, chloride (depending on ash conditions), silicate and, flame tests with optical filters for lithium, sodium, potassium, calcium and copper.

19   Tests on Organics

Separated organics may, after thorough drying, be subjected to Lassaigne's sodium fusion test followed by test tube reactions on the resultant solution: sulfur is indicated by sodium nitroprusside, nitrogen by Prussian blue, chlorine/bromine by silver nitrate.

For the presence of chlorine a very simple test, Beilsteins, is to place a copper wire, with a small amount of substance adhering to it, into a bunsen flame. The appearance of a green flare indicates chlorine.

20   Distillation, Boiling Point/Range

Liquid formulations, particularly those based on solvents, yield useful information when distilled by one of various methods: simple, fractional, steam, azeotropic. Applying these methods to preparations containing surfactants may be made difficult by factors such as foaming and bumping. Although the prudent use of anti-foaming agents and anti-bumping granules may help there is still a need for a considerable degree of care in controlling the rate of boiling.

Simple distillation is normally the first step; where this gives a range of boiling temperatures, indicative of a mixture of volatiles then fractional distillation is called for. Preparations containing substances that would be damaged at high temperatures required for their distillation may be steam distilled. In this procedure the steam is forced through the mixture in the distillation flask and lowers the vapour pressure of the sensitive component which then distils over at a temperature lower than it would normally distil at.

Where an accurate assessment of boiling point/range is required, in simple or fractional distillation, the barometric pressure must be known to correct the observed figures to 760 mmHg.

Azeotropic distillation involves the addition of a volatile liquid to the mixture to be distilled. The choice of liquid is made so as to enable distillation to proceed more effectively than would be the case in the absence of the added liquid. The Dean & Stark method for distilling water out of a substance is just such a method.

21   Dean & Stark

This is specifically for the determination of water content at the few per cent level in liquids and finely divided solids. It involves distilling the sample in the presence of toluene (methylbenzene) that, as it boils at 110 °C, carries over any water present to be collected and measured.

22   Ammonia Distillation

Ammonia, alkylamines and their salts frequently occur as components in formulations. Free ammonia is readily detected by its odour and effect on moist pH indicator paper but not so when its is combined as salts. However, when the salt is treated with alkali, ammonia can be liberated as the gas.

By means of distillation, in which condensate and associated vapour are absorbed in standard acid, followed by titration with standard alkali the free ammonia is determined. Where a similar procedure is applied but with strong alkali added to the sample the amount of combined ammonia can be found.

Ammonia distillation is frequently used in conjunction with Kjeldahl digestion of organics in which organically combined nitrogen is converted into ammonium salts which then go for ammonia distillation.

23   Kjeldahl Nitrogen

The organic matter is digested in boiling concentrated sulfuric acid in the presence of a catalyst. All the organic matter is destroyed and organically-bound nitrogen converted into ammonium sulfate solution that is then used in the ammonia distillation process outlined above. The procedure is nowadays usually carried out in an automated digester to reduce the risk from the high temperature sulfuric acid. Not all types of nitrogen are determinable by this method although modifications may be introduced for certain forms.

24   Sequestering (Chelating) Capacity

The usual procedure for the determination of chelating agents such as EDTA, and one that is suitable for many related compounds, is to titrate an aqueous solution of the sample into a standard solution of calcium chloride in the presence of ammonia buffer and the dye, Eriochrome Black (Solochrome Black).

The EDTA forms a stable complex with the calcium ions until, during titration, it becomes saturated with calcium ions so that further addition produces an excess of calcium which then complexes with the dye to produce a colour change. The method requires quite specific pH levels and determination of the end point may be difficult in all but the simplest – and purest – of samples. Of course being reliant upon a colour change the method is not applicable to coloured samples.

In this case other procedures need be considered such as a potentiometric or polarographic methods. Both these offer much better accuracy and by careful choice of conditions enable one to distinguish between one chelating agent and another.

25   Karl Fisher

Moisture determination in liquids and powders is readily determined by the Karl Fisher titration. Essentially this involves a fairly complex reaction in which iodine is liberated by water present in the sample and is then determined by titration. The traditional apparatus uses a conductivity electrode for end point determination. Small portable kits are available which use a colourimetric end point and syringe/septum bottle for titration.

26   Viscosity

In many formulations there is a need to assess viscosity and flow characteristics to ensure that the product complies with specification or to study how the preparation behaves under different conditions. For liquids and solutions where the mechanism of fluid flow is simple the viscosity may be accurately determined by means of an Ostwald viscometer in which the liquid flows down through a glass capillary tube and is timed between two points. There are many variations of the capillary flow method among which the flow cup is a quick and simple method for used in the mixing plant.

For liquids that do not have simple flow characteristics viscometers based upon rotational methods, rotating cylinders or plates, are used. Here the liquid placed between one or more rotating surfaces is subjected to shearing forces and its resistance to them is measured. Further useful information is obtained by studying how the liquid under test behaves with different shearing rates. This is particularly valuable in the development of preparations that are to be subjected to mechanical action in use or during dispensing from their container. For example non-drip emulsion paints are gels that reduce in viscosity when put under a shearing forces created by the paint brush moving against the gel surface. With such preparations it is essential to study flow characteristics far beyond simple viscosity determinations. The most widely used rotational method for viscosity is the Brookfield viscometer. This instrument enables shearing to be carried out continuously and so allows the study of different conditions and at different shearing rates. Viscosities of suspensions, pastes and gels are commonly determined by means of the Brookfield viscometer.

27   Ross–Miles Foam Test

Determination of the foam-producing capacity of surfactants is normally carried out by means of this test. This is usually carried out on a 0.25% aqueous solution of surfactant, 200 ml of which is allowed to fall from a pipette from a height of 90 cm into a small reservoir of the same solution.

Initial foam height is recorded as soon as the pipette has emptied; foam height after 5 minutes is also taken as an indication of foam stability. A temperature of 60 °C, achieved by thermostatting the apparatus, is usual. Good foaming agents give foam heights of 150 to 200 mm and poor ones are down at 0 to 50 mm.

28   Drave's Test

Primarily for surfactant formulations that are destined for textile use this method assesses the wetting capacity. A skein of standard cotton yarn is attached to a small weight and immersed in a cylinder of the surfactant solution. As the surfactant wets the fibres the air trapped between them is slowly displaced until a point is reached at which the skein suddenly sinks. Wetting effectiveness is taken as the time elapsed between immersion and sinking. A good wetting agent gives a value of up to 10 whereas a poor one may take up to several minutes.

29   Phase Inversion Temperature

Useful in testing of emulsions the phase inversion temperature involves monitoring the conductivity of the emulsion as the temperature is increased at about 2 °C min^{-1}. When a certain temperature is reached the oil-in-water emulsion changes to a water-in-oil emulsion. At this point the water becomes the disperse phase which is indicated by a decrease in conductivity. The temperature at which this occurs is recorded as the phase inversion temperature.

30   Thin-layer Chromatography

A glass or metal plate coated with a fine powder such as alumina of high surface acts as a separation medium. The coating, specifically chosen for the type of compounds to be separated, is referred to as the stationary phase. The sample solution is spotted onto the plate and any carrier solvent is allowed to evaporate, after which the plate is stood vertically in a specially selected solvent mixture that creeps up the plate by capillary action and thus acts as the mobile phase.

As the solvent proceeds vertically it takes with it the substances present in the sample. Each substance experiences a different strength of interaction with the stationary phase and so a separation is effected. Careful choice of stationary phase and mobile phase makes the technique highly effective for separation of both organic substances and inorganic ones. As the solvent front approaches the top of the plate, the plate is removed from the solvent and allowed to dry.

Where the substances are coloured the elution may be assessed visually but most substances are colourless and need to be sprayed with locating agents. In using the technique to assist in identification of the components in a mixture it is common practice to run, simultaneously with the sample, mixtures of known composition.

31   Cloud Point

The clouding of a solution due to the onset of precipitation normally occurs when a solution is cooled but can also occur on heating, as is the case with alcohol ethoxylates where there is a decrease in solubility as temperature increases. In formulations using ethoxylates the cloud point, the temperature at which precipitation and clouding begin, should always be taken into consideration, especially where the formulated product is to be used at elevated temperatures.

32   Saponification Value

Useful when working with compounds containing an ester group, particularly fats and oils, the saponification value involves determining the amount of potassium hydroxide required to saponify (hydrolyse and neutralize) the substance. Saponification is brought about by refluxing the substance with ethanolic potassium hydroxide. Where the substance is difficult to hydrolyse then the ethanol may be replaced with a higher boiling solvent.

INSTRUMENT METHODS

The methods described above often provide the formulator with some basic information relating to composition and many of them are useful as quality control procedures. However, for a full analysis or to get a positive identification of a particular substance, analytical instruments must also be used. Of the various instrumental techniques each one provides a specific type of information – composition or molecular structure – about the substance being studied. In general, instrumental analysis is confined to separated and purified substances.

Only in a few cases would a chemical preparation be placed directly into an instrument. Preparation of the sample for the instrument is an important item and may be very time consuming – especially when compared with the few minutes that the instrument will take to run the sample through.

To filter a sample, precipitate the fatty acids, re-filter, distil the filtrate and then redistill followed by drying the new distillate over anhydrous sodium sulfate for twenty four hours and finally filtering off the liquid seems out of all proportion compared with the twenty minutes it takes to run the sample, along with standards, through the infra-red spectrophotometer and to identify it.

Infra-red (IR) Spectroscopy

The method, which is suitable only for pure substances – and perfectly dry ones at that, relies upon on optical system which passes a beam of monochromatic light, in the region 4000 to 625 nm, through the specially prepared sample followed by detection and measurement of the intensity of transmitted light. The mode of operation is for scanning the region and recording the transmittance spectrum.

The infra-red spectrum so produced will show peaks corresponding to infra-red absorption by the different covalent bonds in the substance and enables the analyst to identify what chemical groups are present along with certain structural features of the substance.

One great asset of infra-red is that each organic compound has its own characteristic absorption spectrum which contains sufficient information for it to be regarded as a fingerprint for that particular compound. Thus, comparison of the spectrum produced from an unknown with that of a known substance enables identification.

Sample preparation always starts with a scrupulously dry material followed by: (a) if the sample is a liquid, forming a thin film between two sodium chloride plates or placing into a sodium chloride cell, or (b) if the sample is a solid, triturating it with potassium bromide and compressing the mixture into a disc or mulling it with liquid paraffin and applying the mull to the sodium chloride plates to form a thin film.

Nuclear Magnetic Resonance (NMR)

Sometimes known as proton magnetic resonance it is used in organic analysis and structure determination. NMR involves preparing the sample as a solution in deutero chloroform along with an internal standard of tetramethylsilane. In the instrument the sample sits in a powerful magnetic field, with a beam of electromagnetic radiation in the frequency range 60–100 MHz passing through it. The radiation absorbed by the hydrogen protons at different field strengths is recorded. A spectrum is thus built up and relates to the structural features of each type of hydrogen atom in the molecule, enabling the molecular structure to be determined.

Mass Spectrometry (MS)

In a mass spectrometer the organic molecules of the vapourized substance are ionized to positive ions by means of electron impact. During this process a particular molecule undergoes partial fragmentation to give a mixture of small ions. These ions are accelerated and focussed by means of an electrodes at high voltage to form an ion beam which then passes through a magnetic field arranged so as to bend the beam.

The effect of this is that the mixture of ions separates depending upon the mass and charge of the ions and a single beam is split into several. The beams pass into a detection system which produces the mass spectrum. A typical molecule gives a molecular ion and several fragmentation ions that are characteristic of that molecule. The information obtained can be compared with that from a computerized data bank and the substance identified.

A recent development to the method for producing the ion beam uses laser light which produces a soft ionization, as compared with the hard ionisation of electron impact. Using the laser method, known as Matrix Assisted Laser Desorption Ionization (MALDI), large molecules such as biomolecules and surfactants can now be analysed by mass spectrometry.

Gas Chromatography (GC)

Gas chromatography is a means for separation and detection of substances in a mixture and when coupled with mass spectrometry it becomes a powerful tool for the identification of the individual compounds. A limiting factor for the GC is that the substance must be volatile (but see Pyrolysis GC) as the technique relies upon the compounds being in the gas phase.

When a substance is not itself volatile it may be possible to make a volatile derivative from it and use that for the GC work. The gas chromatography of fatty acids is an example of derivatisation; it involves reacting the involatile fatty acid with methanol in the presence of boron trifluoride as a catalyst to convert it into the volatile methyl esters of the fatty acids.

The volatile compounds present in the mixture undergoing analysis are injected onto the GC column which contains the stationary liquid phase – usually an involatile wax or oil. This liquid phase is held as a thin film coating on a support powder (packed column GC) or on the inside surface of the column (capillary GC). Upon injection of the volatile sample onto the column the mixture of molecules is carried by means of the carrier gas through the column.

Separation of the different molecules occurs as each type has a different strength of molecular attraction to the liquid phase. This is the basis of the partition effect and its effect upon the moving gas stream is to cause the molecules to separate – in a way, the column exerts a drag on the different compounds. Columns with different liquid phases such as polar ones and non-polar ones may be used to optimize separating power.

At the exit end of the column the separated compounds are eluted one by one and pass into a detection system and recorder which produces the chromatogram showing a series of peaks corresponding to a particular compound and indicating the proportion of it.

Headspace GC

A solid or liquid containing volatile compounds will produce a vapour pressure in the air surrounding the sample. A familiar example is the air freshener gel which releases a mixture of volatile compounds which make up the odour. In headspace GC these volatiles are passed through the GC for analysis. For analytical purposes it is important to ensure that the composition of the volatiles in the gas phases is a proper representation of what was in the solid or liquid.

In practise the usual arrangement is for a small quantity of sample to be heated so that all the volatiles are driven out of the solid or liquid and enter the gas phase. If this is not done to completion then the composition of the gas may not be similar to that in the sample and the final GC results will be wanting. By means of internal standards and reference samples Headspace GC is a valuable and reliable method.

One application of this method is in examining the volatile solvents from the printing ink in chocolate bar wrappers. Although the amount of solvent may be only at the trace level it is essential to study what happens as it evaporates – a small amount of amyl acetate (pentyl ethanoate) migrating into the chocolate and dissolving there in the fats may not be appreciated by the consumer whose bar of whole nut tastes of pear drops.

Pyrolysis GC

In this procedure the sample itself does not pass through the GC column but instead undergoes rapid heating to a temperature at which thermal decomposition occurs. It is the mixture of gaseous products from this reaction that is fed into the column by the carrier gas. From then on the procedure is as for regular GC analysis.

Samples that are involatile or ones that are thermally unstable can provide useful information in Pyrolysis GC. Different compounds give rise to different decomposition profiles. As such, and with strict thermal control to provide reproducible conditions for the thermal decomposition, the technique can provide a fingerprint type chromatogram which is characteristic of a particular substance.

Gas Chromatography–Mass Spectrometry (GC–MS)

The GC techniques outlined above may be used to feed the stream of compounds eluted from the column into a mass spectrometer. This arrangement is a powerful analytical tool and is widely used for the analysis of mixtures of volatile compounds.

Other Instrument Methods

There are other chromatographic methods such as those based upon the liquid phase rather than the gas phase of which High Performance Liquid Chromatography (HPLC) is the main one.

The methods outlined above are, in general, confined to the study of organic molecules. As we have seen most of our everyday formulations are mixtures of organic compounds and so the instrumental methods find wide application in formulation work. However, it is important to recognize that other analytical instruments have a role to play and must be mentioned here. In addition to the organic chemistry of formulations there are also some inorganic aspects, for example many preparations contain salts, mineral acids, acids and alkalis.

There is clearly a need for inorganic analysis. For the determination of metals Flame Photometry (for lithium, sodium, potassium, calcium) and Atomic Absorption Spectrophotometry (for all metals) find wide application whereas for detecting and measuring the amounts of anions the most common approach is to use Ion Chromatography (for phosphate, sulfate, silicate *etc.*).

Bibliography

D. Attwood and A. T. Florence, *Surfactant Systems*, Chapman and Hall, ISBN 0–412–14840–4.

J. H. Clint, *Surfactant Aggregation*, Blackie, ISBN 0–216–92905–9.

E. W. Flick, *Household and Automotive Chemical Specialities*, Noyes Data Corporation, ISBN 0–8155–0751–8.

Handbook of Chemistry and Physics, Chemical Rubber Company, Boca Raton, FL, ISBN 0–8493–0565–9.

G. L. Hollis, *Surfactants Europa*, Tergo Data.

D. C. Callum (Ed.), *Introduction to Surfactant Analysis*, Plenum Publishing Corp, ISBN 0–7514–0025–4.

D. R. Karsa (Ed.), *Industrial Applications of Surfactants IV*, RSC, London, ISBN 0–85186–666–2.

R. G. Laughlin, *The Aqueous Behaviour of Surfactants*, Academic Press, ISBN 0–12–437760–2.

E. Lomax, *Surfactants Encyclopaedia*, Speciality Training Ltd.

G. F. Longman, *The Analysis of Detergents and Detergent Products*, John Wiley, ISBN 0–471–54457–4.

J. L. Lynn and B. H. Bory, *Encyclopaedia of Chemical Technology, Volume 23 (Surfactants)*, Kirk-Othmer, ISBN 0471–52692–4.

B. M. Milwidsky and D. M. Gabriel, *Detergent Analysis*, Halsted Press, ISBN 0–7114–5735–2.

D. Myers, *Surfactant Science and Technology*, ISBN 0–89573–339–0.

M. R. Porter, *Handbook of Surfactants*, Blackie, ISBN 0216–92902–4.

M. J. Rosen, *Structure/Performance Relationships in Surfactants*, American Chemical Society, ISBN 0–8412–0839–5.

D. J. Shaw, *Introduction to Colloid and Surface Chemistry*, Butterworth Heinemann, ISBN 0–7506–1182–0.

Shinoda and Becher, *Principles of Solutions and Solubility*, Marcel Dekker, ISBN 08247–6717–9.

W. Umbach (Ed.), *Cosmetics and Toiletries*, Ellis Harwood Series, ISBN 0–13–181355–2.

Index

abrasive, 46, 122, 123, 124
acetic acid, 21
acetylenic, 75
acid alkyl phosphoric ester, 64
acid dissociation constant, 15
acid salt, 17, 20, 102
acids, 14
acidification, 146
acidity regulator, 99, 101, 102
active matter, 91
N-acyl sarcosines, 45
adsorption, 53, 65, 83, 86
aerosol, 3, 44, 116
aggregate, 76
aggregation, 56, 59
aggregate number, 57
air freshener gel, 119
alcohols, 35, 144
alcohol ethoxylate, 70, 72, 73, 127,
    140
alcohol sulfates, 64
algaecide, 68
aliphatic, 33, 34,
alkalis, 14
alkanolamides, 72
alkyl amido propyl betaine, 108
alkyl aryl sulfonates, 64
alkyl dimethyl benzyl ammonium
    chloride, 68
alkyl glucosides, 73, 76, 117
alkyl phenol ethoxylate, 68
alkyl polyglucosides, 140, 142
alkyl sulfates, 117

alkyl trimethyl ammonium
    methosulfate, 68
alkylbenzene, 139
alkylbenzene sulfonic acid, 139
allantoin, 115, 116
alpha alkene, 138
aluminium chlorhydrate, 111
amalgam, 3
American Society for Testing
    Materials (ASTM), 153
amine ethoxylate, 60, 72
amines, 145
aminocarboxylic acids, 22
2-aminomethylpropan-1-ol, 121
amphiphilic, 53
amphoteric, 14, 60, 73, 76, 93, 106,
    109, 124, 126, 144
ammonia, 3, 8, 18, 119, 122, 137,
    143, 145, 161
ammonium alkyl ether sulfate, 64
    lauroyl sarcosinate, 65
    lauryl ether sulfate, 64, 118
    lauryl sulfate, 107, 108
analysis, 152
anion, 19, 28, 51, 65
anionic, 14, 57, 60, 61, 62, 63, 64, 67,
    68, 74, 85, 93, 94, 103, 106, 109,
    118, 123, 129, 142, 144, 145
anti-cancer agent, 104
antifoam,
antioxidant, 42, 101
antiperspirant, 111
anti-static, 106